POWER ELECTRONICS USING ARDUINO

Jayanand Balakrishnan

SCMS School of Engineering and Technology

Ernakulam

INDIA • SINGAPORE • MALAYSIA

ISBN 979-8-89026-757-3

CONTENTS

PROLOGUE

There are plenty of books on Power Electronics available in the market. All of them are loaded with a lot of theory. This book emphasises the practical implementation of power electronic circuits. Any electric or electronic system consists of circuits. Unlike an ordinary electric or electronic system, a power electronic system consists of two separate entities: the power circuit and the control circuit. Even though the working of any power electronic system requires both parts, most of the textbooks delve deep into the power circuit part only. In this book, more emphasis is given to the digital control part. With a little knowledge of programming, any power electronic system can be implemented using a microcontroller. Arduino Uno board, which is widely used, is used in this book for demonstration.

Power electronic systems have become an integral part of every product existing today. Starting from fans and shaving sets to electric vehicles and power grids. Almost all equipment is becoming smart and intelligent nowadays. Electric lights which were turned ON and OFF through a mechanical switch are now being operated with varied illuminations and colours incorporated in a single lamp with a remote operation facility. Air conditioners and refrigerators are equipped with inverters to operate in an energy-efficient and intelligent manner. Electric vehicle technology has matured and it is expected that all vehicles would be electric within a decade. The share of non-conventional energy sources is growing fast so that only green energy will be prevailing within a couple of decades. Power electronics play a crucial role in this transformation.

Unlike other electric or electronic circuits, power electronic systems are dynamic or live. They require continuous triggering of electronic switches for their functioning. This is analogous to the human body. The human body is live only if the heart beats. The moment the heart stops beating, the human being is dead. Similarly, a power electronic circuit will serve its purpose only when carefully controlled triggering signals are applied at the gates of electronic switches. These triggering pulses are generated digitally with the help of microcontrollers. The signals generated from microcontrollers will not have the power required to trigger a power electronic device. Hence, these signals generated from microcontrollers are to be boosted with the help of appropriate driving circuits. Generally, all these systems will be working in a closed-loop so that the set value of output is tracked always in case of any disturbances or changes in inputs. This necessitates the use of sensors, and the values so sensed will be given to the input pins of the microcontroller. The controller processes these data and gives out appropriate triggering signals to make the output the desired value. All these low-power circuitry, including the microcontroller, are called control circuitry. The other main part, including the power electronic switches that handle a high amount of power, is called the power circuitry.

Power electronic systems are interdisciplinary themselves. In order to understand a power electronic system, one should have a fundamental knowledge of semiconductors, control electronics, control engineering concepts, embedded system design, simulation of systems, thermal engineering, protection systems, inductor design concepts, etc. Some of the preliminary requirements will be covered in this book as and when required.

The brain of any power electronic system is the microcontroller. This produces the triggering pulses equivalent to heartbeats for pumping blood by a heart in a living being. There are many types of microcontrollers available in the market. The family of microcontrollers used for this book is Arduino based. The design of boards and programming environment are open source and are easily available from the net. There are several versions of Arduino boards starting from Arduino Uno. Each Arduino board has its own features and capabilities. Depending upon the complexity of the system we are dealing with, we have to appropriately choose the Arduino board. For a simple buck or boost converter, Arduino Uno would be serving the purpose. If our system involves the implementation of complicated control algorithms, boards like Arduino Due will have to be used. For systems like vector control of motors, digital signal processors (DSPs) will have to be used.

This book is intended for those who do not have any strong programming skills. All the basics needed for programming an Arduino board for a power electronic application are provided in Chapter 2 of this book.

Basic power electronic circuits are classified into the following four categories:

i) DC to DC conversion circuits

ii) DC to AC conversion circuits

iii) AC to DC conversion circuits

iv) AC to AC conversion circuits.

Implementation of all these types of circuits using an Arduino Uno board is discussed in this book. This book can be used as a textbook for teaching a full course on Power Electronics along with practical classes.

01 POWER ELECTRONIC DEVICES

A general power electronic circuit will consist of one or more of the following components.

1. Switches
2. Inductors
3. Capacitors
4. Resistors
5. Diodes

Out of these components, power electronic switches are the heart of any power electronic system. The heart of any living being beats continuously and the moment the beating stops, the living being dies. Similarly, for a power electronic system to function in a particular manner, firing pulses or gate signals are to be applied to the gates of the switches in a very carefully controlled manner. Once these pulses go awry, the power electronic system is as good as dead.

These switches can be of any type, like Silicon Controlled Rectifiers (SCR), Metal Oxide Semiconductor Field Effect Transistors (MOSFET) and Insulated Gate Bipolar Transistors (IGBTs). These devices are characterised by different turn-on and turn-off mechanisms.

A diode gets automatically ON whenever the voltage across it becomes positive. It automatically turns OFF whenever the voltage across it becomes negative. It means that the circuit conditions decide the state of a diode. Hence, a diode is called a passive switch. SCR and other power electronic switches have the ability to block positive voltages across them. They will start conduction only when the triggering signal is applied to their gate or base terminals. Hence, these switches are called active switches.

1.1 Silicon-Controlled Rectifiers (SCRs)

SCR was invented in 1957 by a team of scientists at Bell Laboratories. The era of power electronics began with the invention of SCRs SCRs were developed as PNPN junctions which were able to withstand very large voltages in the forward- and reverse-biased conditions, contrary to ordinary transistors that existed during that period, which could withstand only small voltages.

SCRs are controllable diodes. An SCR is turned ON by applying a pulse at the gate terminal. Once it is turned on, we lose control over it. It remains ON until the current through it is brought back to zero. In the case of AC circuits, there will be natural zeros in every cycle of the sine wave so that the SCR will automatically be turned off when the current becomes zero. One has to keep on applying pulses at the appropriate points in the positive half-cycle to make the SCR conducting in every cycle, continuously. In the case of DC circuits, once it is turned ON, it can be switched off only after bringing the current to zero by passing an equal amount of current in the opposite direction with the help of an auxiliary circuit called a commutating circuit. The commutating circuit will consist of energy storage elements like capacitors or inductors, capable of providing discharge currents in a direction opposite to the load current carried by the switch.

Even though an SCR will be represented by a simple symbol with three terminals for all power electronic circuits, as shown in **Fig. 1.1**, an actual SCR will have a gate-triggering circuit and/or a commutating circuit associated with it, which will not be shown in circuit diagrams. The control pulses will generally be generated by a low-voltage electronic control circuitry or digitally by a microcontroller/digital signal processor/FPGA.

▲ **Fig. 1.1:** Circuit symbol of an SCR

The control pulses generated from a microcontroller cannot directly be applied to the gate of the switches due to two reasons. The first reason is that the voltages generated at the output ports of a processor at a voltage level of either 3.3 V or 5 V will not have the driving capacity to provide the charging currents needed for making the device fully in the turn ON state by driving the device into full saturation. Hence, the power level of the signal has to be stepped up. The second reason is that proper isolation between the control and power circuit is to be provided for every switch. The triggering pulses are applied across the Gate and Cathode terminals. Since the Cathode terminal is part of the power circuit, if the control pulse is directly applied across the Gate and Cathode, the Cathode terminal and the Ground terminal of the control circuitry will be short-circuited. Moreover, if there is more than one switch in a circuit, all the Cathode terminals will get short-circuited. Hence, proper isolation between power and control circuits must be provided. Isolation is generally provided either by an optocoupler or through a pulse transformer. A typical triggering circuit, used throughout this book as gate drivers for SCRs and IGBTs, is shown in **Fig. 1.2**.

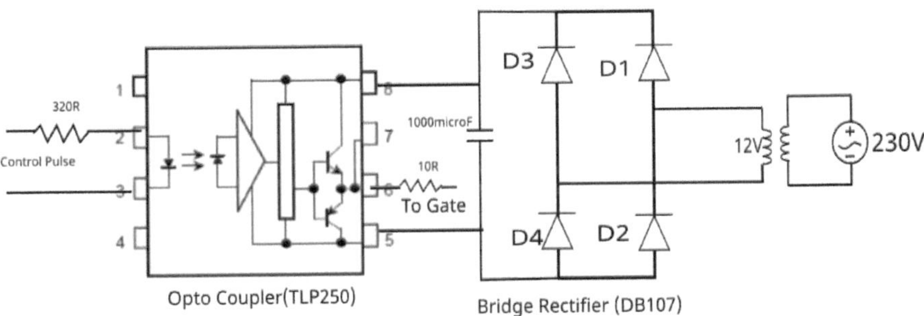

▲ **Fig. 1.2:** Gate Driver Circuit

In order to avoid short-circuiting different Cathode terminals of switches, individual transformers to step down 230 V to 12 V must be used for every switch. A bridge rectifier (DB107) converts this 12 V AC into a DC voltage, along with a filter capacitor. This isolated DC voltage becomes the supply voltage for the optocoupler (TLP250). The thyristors for which their Cathode terminals are connected in the power circuit, the same isolation transformer can be used.

The main disadvantage of SCRs is the fact that they cannot be turned off at our will. Once it is turned on, turning it off is possible only by making the current reach zero value. Hence, one has to wait until the natural zero point of current in the case of AC input. In the case of DC input, one has to provide auxiliary circuits involving other thyristors and passive components like inductors and capacitors for forcing a current in the opposite direction to make the current through the device equal to zero. This increases the size and complexity of SCR circuits. This compelled scientists to continue their search for other types of power electronic switches.

1.2 Gate Turn-Off Thyristors

The main difficulty with thyristors was the difficulty in turning them OFF. This difficulty was overcome partially by the invention of GATE turn-off thyristors (GTOs). A GTO can be turned off by applying a negative pulse at its gate terminal. The difficulty with this turn-off mechanism is the requirement of separate control circuits for turning ON and turning OFF. The circuit representation of GTO is using the symbol shown below:

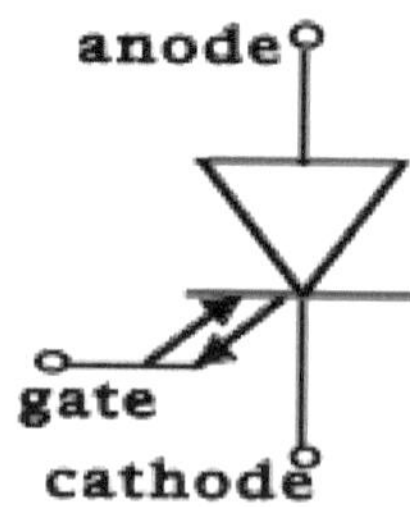

▲ **Fig. 1.3:** Symbol of GTO

1.3 Power Transistors

By changing the doping levels of ordinary transistors, it became possible to enhance the voltage and current handling capabilities of ordinary transistors. The development of power transistors opened up a new era in the control circuits in power electronics. It became possible to turn ON and turn OFF a power electronic switch by controlling the same gate signal itself. A power transistor is ON when the base current, sufficient enough to bring it into the condition of saturation, is applied. It remains ON until this base current is present and it becomes OFF the moment this base current is removed. Generation of control signals became easy with this type of turn-on and turn-off mechanisms. But, the disadvantage was that since transistors are basically current-driven devices, the base current needed for turning them ON is decided by the current gain, β of transistors. For low-current transistors, this current gain will be around 100. But, for power transistors, the value of β is as low as 10. Hence, in order to control a transistor carrying a load current of 100A, a base current of 10A will be needed. Control circuits capable of carrying this amount of current would not only be bulky, but the power requirements of the control circuit would also be high. Hence, power transistors are not very popular in power electronics. The circuit symbol of a power transistor is the same as that of ordinary transistors.

Efforts of scientists to find a solution to this issue culminated in bringing the concept of field effect transistors into the power electronic field. This resulted in the invention of power MOSFETs.

1.4 Power MOSFETs

The main advantage of FET is that it is a voltage-driven device rather than a current-driven device. It can be brought into full saturation by applying a sufficient voltage at its gate terminal. The gate current of FETs will be negligible. This fact made the power requirements of control circuitry, negligible. Moreover, the frequency of switching could also be increased to almost 100 kHz with MOSFETs. This is because FETs are unipolar conduction devices. The current is constituted by either holes or electrons only. This eliminates the existence of reverse recovery charges while turning off and subsequent events which result in slow turn-on and turn-off. An increase in switching frequency means a proportional reduction in the requirement of filter components. MOSFETs are presented in **Fig. 1.4**.

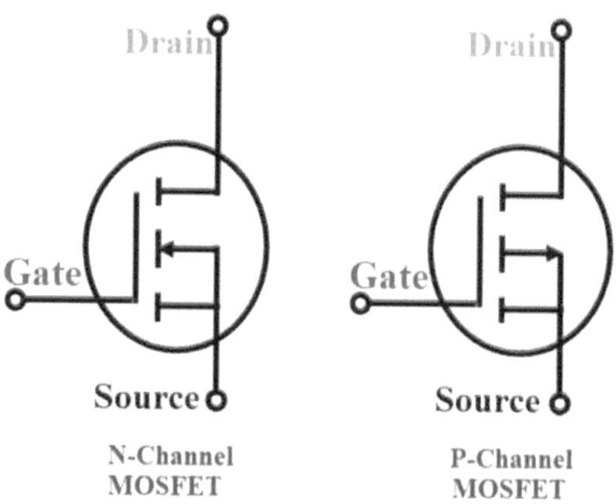

▲ **Fig. 1.4:** Symbols of MOSFETs

The main disadvantage of MOSFETs was their high ON-state resistance. At full loads, there will be a relatively large voltage drop across this resistance, leading to large conduction loss and poor efficiency. This difficulty was overcome by scientists through the invention of another device called an IGBT.

1.5 Insulated Gate Bipolar Transistor

IGBTs combine the good qualities of both power transistors and MOSFETs. The gate side is similar to the gate of a MOSFET. It is a voltage-driven gate so that the gate current needed for turning on is very less when compared to that of a power transistor. The main conducting path is similar to that of a transistor. This results in low ON-sate resistance leading to a reduction in conduction loss. Hence, the main conducting terminals are designated as collector and emitter as in the case of transistors and the base terminal is designated as Gate, as with a MOSFET. Most of the medium-power applications in the field of power electronics use IGBTs as switches. IGBTs are represented using the symbol shown in **Fig. 1.5**.

▲ **Fig. 1.5:** Symbol of IGBT

Even though the principle of operation of turn on and turn off of IGBTs are similar to those of MOSFETs, the gate-to-emitter capacitance of IGBTs is relatively high. The gate signal takes some rising time to appear between the gate and emitter terminals due to this large capacitance. Hence, the maximum switching frequency possible with IGBTs is slightly less than that with MOSFETs. The gate driver circuit shown in **Fig. 1.2** can be used for IGBTs for operating frequencies up to a few kilohertz. For higher switching frequencies, special driver circuits having the capability to force in large charging currents are to be used. A negative voltage is provided at the gate for a faster turn-off process. Standard driving boards are available in the market for faster operations.

IGBTs are manufactured for standard voltage ratings of 600 V, 1200 V and 1700 V. SCRs are still being used for higher voltage and power levels.

1.6 Wide Band Gap Devices

All the power electronic switches mentioned in earlier sections are silicon devices. The voltage and current ratings of these devices got increased stage by stage. Scientists have met with a deadlock on further enhancement of voltage and current ratings. They have realised that further enhancement in ratings is not possible with silicon as the material to make these devices. This is because of the fact that silicon has reached its theoretical limit on its ratings. Further enhancement is not theoretically possible with silicon. Scientists have searched for new devices to make power electronic switches and the result was the development of wide band gap (WBG) devices. Silicon Carbide (SiC) and Gallium Nitride(GaN) are the most popular WBG devices.

The voltage rating of a device depends on the capacity of the material to withstand voltages without breakdown. The breakdown voltage depends on the gap between the valence and conduction bands of the material. The band gap is the difference in energy levels of the valence and conduction bands. In metals, valence and conduction bands are interleaved and electrons are available for conduction under normal conditions, and hence they have large conductivity values. On the other hand, insulators have a wide energy gap between the conduction and valence bands (**Fig. 1.6**). Hence, huge energy is required to bring electrons from the valence band to the conduction band. That is why they have poor conductivity. Semiconductors have moderate values of band gap so that under normal conditions, some electrons are available in the conduction band and they contribute to intrinsic conductivity. When doped with appropriate materials or with an increase in temperature, more electrons and holes are created for increased conductivity. This moderate band gap limits the capability of silicon devices to withstand large voltages before getting broken down. SiC and GaN have larger band gaps than that of Si and hence they are called WBG devices.

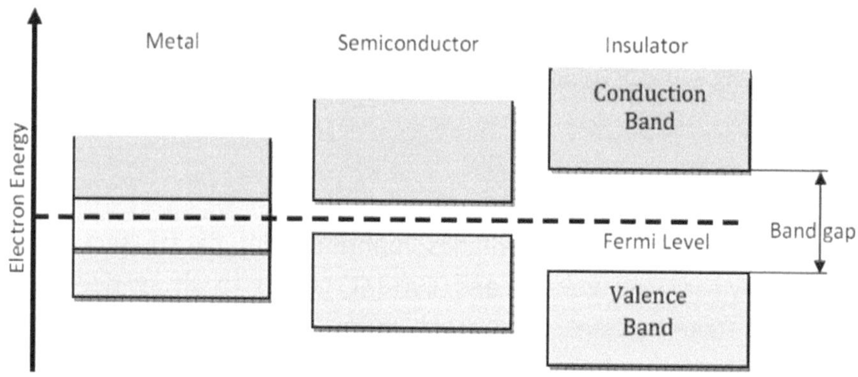

▲ **Fig. 1.6:** Concept of band gap

The band gap is specified using the unit of electron volts (eV). For silicon and germanium, band gaps are 1.1 eV and 0.7 eV, respectively. For the new-generation power electronic materials like SiC and GaN, they are 3.3 eV and 3.4 eV, respectively. Hence, power electronic switches capable of blocking hundreds of kilovolts are possible to be made using these devices.

All the disadvantages of ordinary switches are addressed in SiC and GaN devices. These devices have reduced size, reduced ON-state resistance and reduced gate capacitance. Since losses are less, the size of the heat sink needed is very less. It is possible to combine the power circuit as well as the control circuit in the same printed circuit board. At present, a motor drive system is bigger than the motor which is being controlled. A WBG-based drive system can be embedded in the motor itself.

The extent of size reduction and ON-state resistance reduction is depicted in **Fig. 1.7** in which the size and resistance of a Si device and SiC device are being compared.

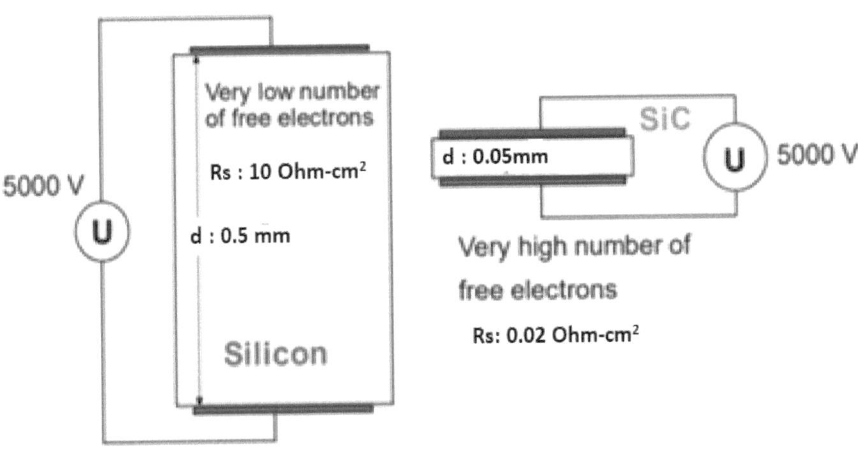

▲ **Fig. 1.7. Comparison of size and resistance of Si and SiC devices**

A Si device and a SiC device having the same voltage rating of 5000 V are compared. The thickness of the Si device is 0.5 mm and that of the SiC device is 0.05 mm. It means a 10-fold reduction in size. At the same time, the specific resistance of SiC is getting reduced by a factor of 500. This means a proportional reduction in losses and heat sink requirement.

The junction capacitance of SiC devices is negligible so high-frequency operation to the tune of mega Hertz is possible. This means very low values for filter inductance and capacitance.

02 ARDUINO PROGRAMMING

2.1 The Arduino Family

Arduino is an open-source electronics platform based on easy-to-use hardware and software. It consists of a series of microcontrollers on a single integrated circuit that can be programmed to perform a wide range of tasks. There are many different types of Arduino boards available, each with different specifications and capabilities. One of the key features of Arduino is its vast community of users who share their projects, code and knowledge online. This makes it easy for beginners to learn and for experienced users to collaborate and innovate.

The main members of the family are listed below:

- **Arduino Uno:** This is the most popular Arduino board and is widely used in various projects. It is based on the ATmega328P microcontroller and has 14 digital input/output pins, 6 analogue input pins, a 16 MHz quartz crystal, and a USB connection for programming and communication with a computer.

- **Arduino Mega 2560:** This board is similar to the Uno but has more input/output pins, making it ideal for larger projects. It is based on the ATmega2560 microcontroller and has 54 digital input/output pins, 16 analogue input pins, and 4 hardware serial ports.

- **Arduino Nano:** This is a small board that is similar to the Uno but in a more compact form. It is based on the ATmega328P microcontroller and has 14 digital input/output pins, 8 analogue input pins, and a mini-USB connection for programming and communication.

- **Arduino Due:** This board is based on the Atmel SAM3X8E ARM Cortex-M3 CPU and has 54 digital input/output pins, 12 analogue input pins, and 2 digital-to-analogue converters (DACs). It also has a USB connection for programming and communication.
- **Arduino Leonardo:** This board is based on the ATmega32U4 microcontroller and has 20 digital input/output pins, 12 analogue input pins, and a built-in USB connection for programming and communication. It also supports mouse and keyboard emulation.
- **Arduino Pro Mini:** This is a small, low-cost board similar to the Nano but without the USB connection. It is based on the ATmega328P microcontroller and has 14 digital input/output pins and

Arduino Uno is extensively used in this book for implementing power electronic circuits. The pin diagram of the Arduino Uno board is shown in **Fig. 2.1**

The main features of Arduino Uno are mentioned below:

- **Microcontroller:** The Arduino Uno is based on the ATmega328P microcontroller, which has 32KB of flash memory, 2KB of SRAM, and 1KB of EEPROM.
- **Digital I/O pins:** The board has 14 digital input/output pins, which can be used as digital input or output. These pins operate at 5 volts.
- Arduino Uno has 6 analogue input pins, which can also be used as digital input/output pins. The analogue pins can measure voltage between 0 and 5 volts.
- The board has 6 PWM (Pulse Width Modulation) pins, which can be used to generate analogue output signals.
- The board has a USB port and a TTL serial port, which can be used for serial communication with a computer or other devices.

- The board can be powered through a USB port or a 9 V battery, and it can provide up to 20 mA of current to external devices.
- **Integrated Development Environment (IDE):** The Arduino IDE is a simple and easy-to-use software development environment that allows users to programme the board using the C++ programming language.
- **Open-source hardware:** The Arduino Uno is an open-source hardware platform, which means that the board's schematics and design files are freely available for anyone to use and modify.

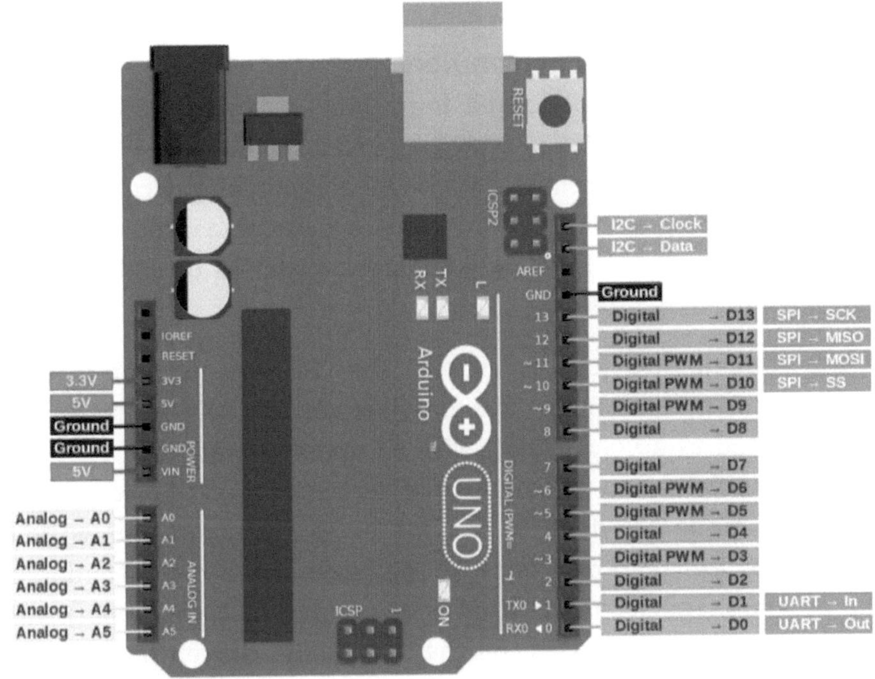

▲ **Fig. 2.1:** Pin diagram of Arduino Uno Board

2.2 Features of Arduino Useful for Power Electronic Systems

In a power electronic system, one has to measure the voltages/currents/ speed of the system and convert them to the levels acceptable to the processor boards. Arduino Uno board, which is used in this book for demonstration, can accept analogue inputs in the range of 0 V to 5 V. Hence, all AC signals are to be level shifted by 2.5 V with a peak value of 2.5 V. After processing these input signals, the output will be the triggering signals for IGBTs or thyristors. These signals are outputted through digital input/output pins. The digital input and output pins can handle voltages of either 0 V or 5 V.

2.3 Programming Environment

Arduino IDE can be downloaded from: https://support.Arduino.cc/hc/en-us/articles/360019833020-Download-and-install-Arduino-IDE. This provides an easy and excellent platform for programming microcontroller boards. This programming environment allows for a simple way for programming, without knowing any hardware details or addresses for data, memory or registers.

If the Arduino IDE icon is clicked, the IDE gets opened with a template for programming. The template consists of two parts. One is designated as a void setup() with an opening brace and a closing brace. All initialisation commands like the declaration of pins as input or output mode are done in this space. All the commands written here will be executed only once, at the beginning of the execution of the program. The second portion is designated as a void loop() with opening and closing braces. All the commands written here will be repeated indefinitely.

Some of the basic functions used in Arduino programming for power electronic circuits are described below:

2.4 Analogue Input

The function to be used for reading voltages or currents is **analogRead(pin).** Useable pins in an Uno board are A0 to A5, with a resolution of 10 bits. An input voltage of 0 V will be read as 0 and a voltage of 5 V will be read as 1023. An input voltage of 2.5 will be read as 255. The following example reads a signal at pin A0 and displays it on the monitor continuously.

```
int pin = A0; // variable name for analogue input pin
int val =0;   // variable name for storing values of ADC output


void setup() {
Serial.begin(9600); // baud rate for sending data to serial monitor
}


void loop() {
val=analogRead(pin); // reading value at A0
Serial.println(val); // sending data to serial monitor
}
```

The fixed ends of a potentiometer can be connected at Vcc and ground pins of an Arduino board and the variable point to A0. Download the programme to the Arduino board. Click on the serial monitor icon at the top right corner of the Arduino IDE. The value of the variable read through the analogue pin will be displayed on the screen. The change in values depending upon the change in the position of the variable point in the potentiometer can be observed on the monitor.

2.5 Digital Output

After reading voltages and current values, we will be writing codes for processing these signals. The output from the processor will be the gating pulse signals to be given to power electronic switches. These are logic signals and are treated as digital signals in Arduino. In Arduino UNO, digital signals will have a voltage level of 5 V. The command for digital output is **digitalWrite(pin,VALUE).**

The following programme segment shows how to send logic signals at the digital pin 3:

```
void setup(){
pinMode (3,OUTPUT); // Declares pin 3 as output port
}

void loop() {
digitalWrite(3,HiGH); // puts logic 1 signal at pin 3
delay(1000);   // waits for 1 second
digitalWrite(3,LOW);  // puts logic signal 0 at pin 3
delay(1000);
}
```

2.6 PWM Generation

For the operation of inverters and converters, one has to dive a train of pulses with the required duty cycle to the gate of IGBTs or MOSFETs. These signals are called pulse width modulated waves. Some of the digital pins of Arduino boards are capable of providing pulse width modulated signals directly.

In Arduino UNO, there are six pins that support PWM output. These pins are marked with a "~" symbol on the board and are as follows:

- Pin 3
- Pin 5
- Pin 6
- Pin 9
- Pin 10
- Pin 11

One can use these pins for speed control of motors and the brightness of lights, among other things. The PWM technique works by varying the duty cycle of the square wave signal output by these pins, which in turn varies the average voltage or current delivered to the attached device.

One can generate PWM signals using the analogWrite() function. The following steps are needed for using the PWM pins:

Set the output mode of a PWM pin using the pinMode() function:

```
pinMode(3,OUTPUT); // sets digital pin 3 in output mode
Generate PWM signals at pin 3 using the analogWrite() function:
analogWrite(3,125); // generates PWM signal at pin 3 with 50% duty cycle
```

Pulse width can be mentioned as an integer number varying from 0 to 255. 0 corresponds to 0 duty cycle and 255 correspond to 100% duty cycle.

The default frequency of PWM signals is 490Hz at pin 3. This can be made to any other desired frequency by installing the PWM library. This can be installed from the Library Manager in the Tools menu of the IDE of Arduino. After installing this library, the frequency of PWM can be changed to any value by executing the following functions:

```
#include <PWM.h>
```

and

```
SetPinFrequencySafe(3, 1000); // set 3rd pin frequency as 1 kHz:
```

2.7 Setting the Sampling Period

Practically, almost all power electronic systems will be working in a closed loop. Output parameters, like the speed of a motor or the voltage of a converter, are to be maintained constant using a controller algorithm, which has to be executed repeatedly at a fixed sampling rate. The controller parameters would be designed for a particular sampling frequency and the digital controller has to ensure the execution of this algorithm with a fixed sampling rate. This can be done using **Timers** in Arduino. All control algorithms will be executed as Interrupt Service

Routines (ISRs). The sampling time is set by setting the timer period for the execution of the interrupt service routine. This is achieved by installing the **TimerOne** library. The execution frequency can be set using the following commands:

```
#include <TimerOne.h>
Timer1.initialise(1000); // sets sampling frequency as 1000 microseconds
Timer1.attachInterrupt(Name); // executes the ISR specified using Name
```

2.8 Analogue Output

Some of the Arduino boards like Arduino Due have a built-in analogue-to-digital converter (ADC) in the processor itself. In Arduino Uno, there is no direct ADC pin. However, we can use PWM pins to simulate the ADC output. If the PWM signal is passed through a low pass filter, an almost constant DC voltage varying from 0 to 5 V can be obtained from a PWM pin by varying duty cycles from 0 to 255.

2.9 Use of Arduino Serial Port for Troubleshooting

It is often required to troubleshoot a programme using observing the values of a parameter within a programme. One can display the value of any parameter in the computer monitor itself using the serial port facility in Arduino. The following initialisation command is used for setting the baud rate for data transfer and to display the value of a variable.

```
Serial.begin(9600); // sets the baud rate as 9600
Serial.println(variable_Name); // displays the value of variable in
the monitor
```

The serial monitor can be viewed by clicking the **Serial Monitor** in the **Tools** menu of Arduino IDE. It is possible to obtain a time plot of this variable by choosing the **Serial Plotter** in the **Tools** menu.

2.10 Time-related Functions

The following time-related functions will be useful while writing Arduino programmes for power electronic systems:

a) delay()

This function is used for inserting a delay specified in milliseconds in the programme. This function helps create square waves of triggering pulses of any duty cycle. The sample code segment written below can be used for a square wave generation of 1 Hz.

```
void loop() {
digitalWrite(3,HIGH); // makes the pin 3 HIGH
delay(500); // waits for 500 milliseconds
digitalWrite(3,LOW); // makes the pin 3 LOW
delay(500);    // waits for 500 milliseconds.
}
```

b) delayMicroseconds()

This function is used for inserting a delay specified in microseconds in the programme. The sample code segment written below can be used for a square wave generation of 1 kHz.

```
digitalWrite(3,HIGH); // makes the pin 3 HIGH
delay(500); // waits for 500 microseconds
digitalWrite(3,LOW); // makes the pin 3 LOW
delay(500);    // waits for 500 microseconds.
```

2.11 Displaying Values Using LCD

In order to display results an intermediate value of variables for troubleshooting or for displaying measured values of currents, voltages or speeds, liquid crystal displays (LCDs) are used. An I2C adapter board will also be attached to LCDs so that only two pins of Arduino Uno are needed for displaying data in LCD using serial communication. The pins are SDA and SCL. A typical LCD along with an I2C controller is shown in **Fig. 2.2**.

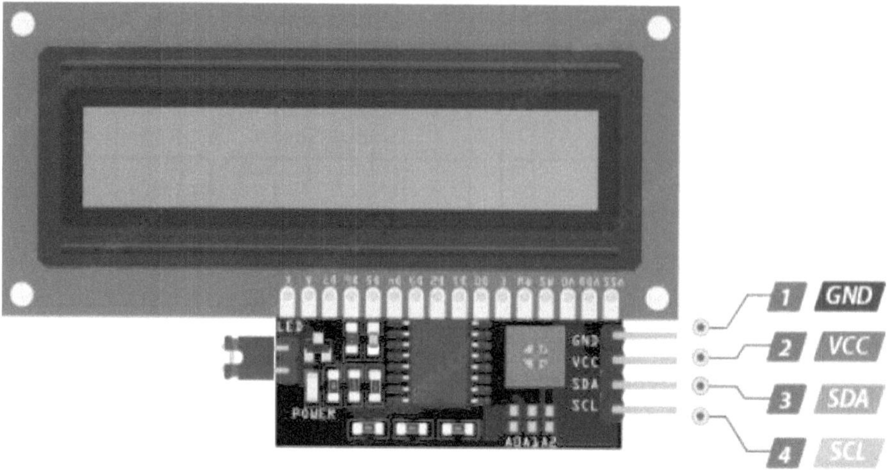

▲ **Fig. 2.2:** LCD connected with I2C controller

In order to use an LCD along with Arduino, **LiquidCrystla_I2C** library has to be installed. A typical programme for displaying 'Hello World' in the display is shown below:

```
#include <LiquidCrystal_I2C.h>
Void setup() {
LiquidCrystal_I2C lcd(0x27,16,2); // specifying the address (0x27) and
size (16 by 2) of display)
lcd.init();
lcd.clear();
lcd.backlight();
lcd.setCursor(2,1);
lcd.print("Hello World");
}
Void loop(){
}
```

2.12 Mathematical Operations in Arduino Uno

The Arduino Uno microcontroller supports various mathematical operations that can be used in programming. Here are some of the most common mathematical operations that can be performed:

- **Addition (+):** The addition operator is used to add two values together. For example, int sum = 3 + 5 will assign the value of 8 to the variable 'sum'.
- **Subtraction (–):** The subtraction operator is used to subtract one value from another. For example, int difference = 10–3 will assign the value of 7 to the variable 'difference'.
- **Multiplication (*):** The multiplication operator is used to multiply two values together. For example, int product = 2 * 6 will assign the value of 12 to the variable 'product'.
- **Division (/):** The division operator is used to divide one value by another. For example, int quotient = 10 / 2 will assign the value of 5 to the variable 'quotient'.
- **Modulo (%):** The modulo operator is used to find the remainder of a division operation. For example, int remainder = 10% 3 will assign the value of 1 to the variable 'remainder'.
- **Increment (++):** The increment operator is used to increase the value of a variable by 1. For example, int count = 0; count++ will assign the value of 1 to the variable 'count'.
- **Decrement (—):** The decrement operator is used to decrease the value of a variable by 1. For example, int count = 10; count— will assign the value of 9 to the variable 'count'.

In addition to these basic mathematical operations, the Arduino Uno also supports advanced mathematical functions such as trigonometric functions (sin, cos, tan), logarithmic functions (log, exp), and square root (sqrt). These functions are part of the standard C++ math library and can be used in Arduino programming.

2.13 Logical Operations

The Arduino Uno microcontroller also supports various logical operations that can be used in programming. Here are some of the most common logical operations that can be performed:

- **Logical AND (&&):** The logical AND operator is used to check if both operands are true. For example, if (x > 0 && y < 10) will evaluate to true only if x is greater than 0 and y is less than 10.

- **Logical OR (||):** The logical OR operator is used to check if either operand is true. For example, if (x == 0 || y == 0) will evaluate to true if either x or y is equal to 0.

- **Logical NOT (!):** The logical NOT operator is used to invert the truth value of an expression. For example, if (!(x > 0)) will evaluate to true if x is less than or equal to 0.

- **Bitwise AND (&):** The bitwise AND operator performs the AND operation on each bit of two integers. For example, int result = a & b will assign the result of the bitwise AND operation between a and b to the variable 'result'.

- **Bitwise OR (|):** The bitwise OR operator performs the OR operation on each bit of two integers. For example, int result = a | b will assign the result of the bitwise OR operation between a and b to the variable 'result'.

- **Bitwise XOR (^):** The bitwise XOR operator performs the exclusive OR operation on each bit of two integers. For example, int result = a ^ b will assign the result of the bitwise XOR operation between a and b to the variable 'result'.

- **Bitwise NOT (~):** The bitwise NOT operator inverts all bits of an integer. For example, int result = ~a will assign the result of the bitwise NOT operation on a to the variable 'result'.

These logical operations can be used in conditional statements, loops, and other programming constructs to implement complex logic and decision-making in Arduino programming.

2.14 Conditional Statements and Loops

The Arduino Uno microcontroller supports various conditional statements and loops that can be used in programming. Here are some of the most common ones:

1. **if statement:** The if statement is used to test a condition and execute code if the condition is true. For example:

```
if (x > 0) {
// do something if x is greater than 0
}
```

2. **if-else statement**: The if-else statement is used to test a condition and execute different codes depending on whether the condition is true or false. For example:

```
if (x > 0) {
// do something if x is greater than 0
} else {
// do something else if x is not greater than 0
}
```

3. **switch statement**: The switch statement is used to test a variable against a list of values and execute different codes depending on the value of the variable. For example:

```
switch (x) {
case 1:
// do something if x is 1
break;
case 2:
// do something if x is 2
break;
default:
// do something else if x is not 1 or 2
break;
}
```

4. **for loop:** The for loop is used to execute a block of code a specific number of times. For example:

```
for (int i = 0; i < 10; i++) {
// do something 10 times
}
```

5. **while loop:** The while loop is used to execute a block of code while a condition is true. For example:

```
while (x > 0) {
// do something while x is greater than 0
}
```

03 DC–DC CONVERTERS

DC–DC converters convert a DC voltage level to another DC voltage level. They have become an integral part of almost all power electronic systems, including renewable power generation and electric vehicles. Since the input here is DC, control would be easy if the switching device is an IGBT or a MOSFET. Even though thyristors can be used for this purpose, an additional commutation circuit will also be incorporated for the turning-off process. This will make the system bulky. Moreover, fast switching operation is not possible with thyristors. In DC to DC converters, the switching frequency has to be to the tune of even hundreds of kilohertz to reduce the size of inductors and capacitors as well as to keep the electromagnetic interference at bay. These are possible with self-turn-off devices like IGBTs and MOSFETs.

DC to DC converters are broadly classified into three categories. They are

a) Buck Converters in which the output voltage is always less than the input voltage.

b) Boost Converters in which the output voltage is always greater than the input voltage.

c) Buck-Boost Converters in which the output voltage will be less than the input voltage for a certain range of duty cycle and it would be greater than the input voltage for the other range of duty cycle.

Design and control of all these three categories are almost the same. Hence, the theory, design, simulation and control of only the buck converter are described at length here. The same control strategies can easily be extended to other circuits.

Depending on the circuit topologies, DC to DC converters are also classified into other categories such as Ćuk converters, flyback converters, resonant converters etc. However, all these converters are controlled by triggering the switching devices at a specific duty cycle. Hence, the programmes for controlling all these circuits remain the same.

3.1 Buck Converter

A buck converter converts a fixed DC input voltage to a variable DC output voltage lower than the input voltage. The circuit diagram of a simple buck converter is shown in **Fig. 3.1**.

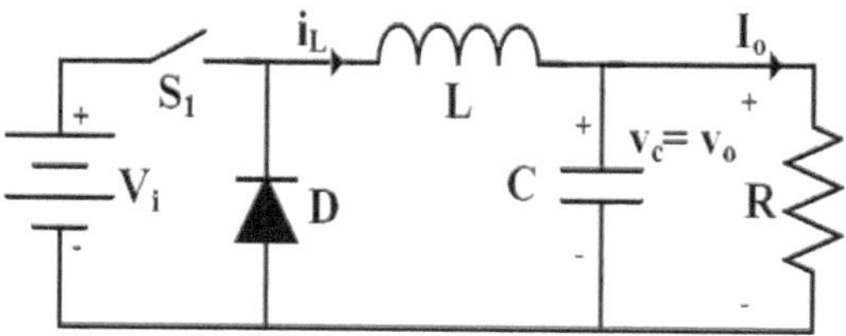

▲ **Fig. 3.1:** Buck Converter

The output voltage from a buck converter can be expressed as

$$v_o = V_i * d \tag{3.1}$$

where V_o, V_i and d are variable output voltage, source voltage and duty cycle ($t_{ON}/t_{ON}+T_{OFF}$), respectively. By varying the duty cycle of the applied pulses to the semiconductor switch, the output voltage can be varied. There are two modes of operation for this buck converter. The first mode is when the switch S1 is ON. The current flow during this mode is depicted in **Fig. 3.2**. The inductor and capacitor store energy during this period. When the switch is OFF, the flow of current will be as shown in **Fig. 3.3**. During this period, the inductor and capacitor discharge their stored energy.

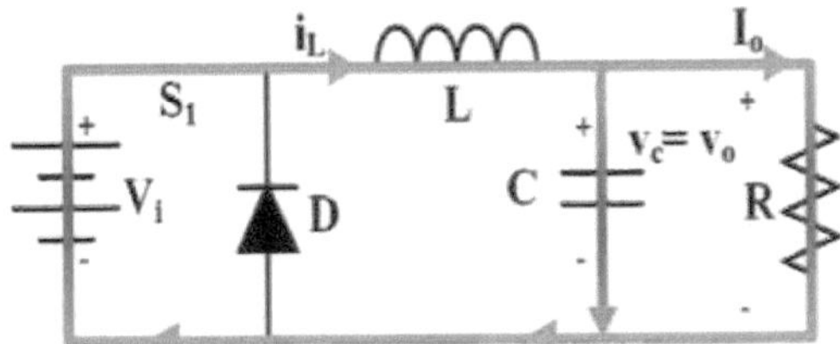

▲ **Fig. 3.2:** Current flow when S1 is ON

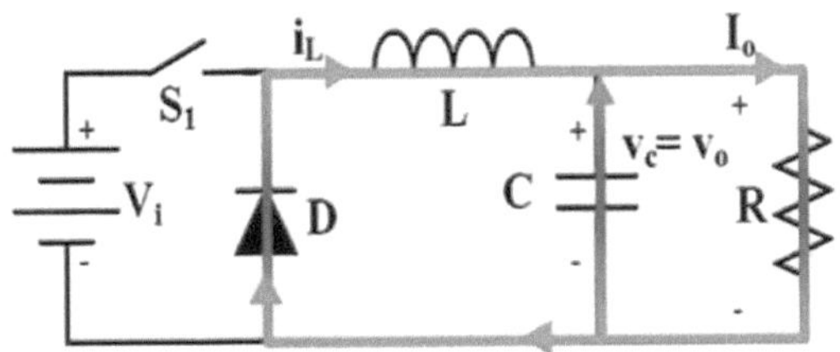

▲ **Fig. 3.3:** Current flow when the switch is OFF

Necessary equations can be derived using the volt-second balance equation across the inductor. As per the volt-second balance principle, the sum of the product of average voltage across the inductor and time for which the switch is ON and the product of voltage and duration for which S1 is OFF will be zero. The voltage across the inductor when the switch is ON will be (V_1–v_o) and the voltage when S1 is off will be –v_o. The duration for which the switch remains ON will be dT and the duration for which S1 is OFF will be (1–d)T, where T is the time period for a cycle of switching.

$$(V_1 - v_o) * dT - v_o * (1-d)T = 0 \tag{3.2}$$

from which we get equation 3.1.

$$v_o = V_1 * d$$

3.2 Mathematical Modelling Of A Buck Converter

In order to design a controller for a buck converter for closed-loop operation, it would be better to derive a mathematical model for the converter. The model can be either in the form of state space representation or in the form of a transfer function. The current through the inductor and the voltage across the capacitor are taken as state variables. Since we have two modes of operation for this circuit, we have to find the time-averaged values for these quantities.

a. Current through the inductor

The voltage across the inductor when the switch is ON is V_1–V_o and it is –V_o when the switch is OFF. The output voltage is the same as the voltage across the capacitor, V_c. The switch is ON for a period of dT and it is OFF for a period of (1–d)T. Hence, the time-averaged current through the inductor can be represented as

$$\frac{di_L}{dt}=\frac{1}{L}\left[(V_i-v_c)d+(-V_c)(1-d)\right] \tag{3.3}$$

b. Voltage across the capacitor

The derivative of the voltage across a capacitor is 1/C times the current through it. Here, the current through the capacitor is the difference between the inductor current and the current through the resistive load.

Similarly, the time-averaged voltage across the capacitor can be expressed as

$$\frac{dv_c}{dt}=\frac{1}{C}\left[\left(I_L-\frac{v_C}{R}\right)dT+\left(i_L-\frac{v_C}{R}\right)(1-d)T \tag{3.4}$$

Simplifying these two equations, we get the two dynamic equations for the converter as

$$\frac{di_L}{dt} = \frac{-v_C}{L} + \frac{V_i d}{L} \tag{3.5}$$

$$\frac{dv_C}{dt} = \frac{i_L}{C} - \frac{v_C}{RC} \tag{3.6}$$

These two equations form the state equations for this converter. The two states are i_L and v_C. The input is the duty cycle d and the output is $v_{C.}$

In order to find the transfer function of the system, apply Laplace Transform to both sides of the state equations:

$$sI_L(s) = -\frac{1}{L} V_C(s) + \frac{V_i}{L} d(s) \tag{3.7}$$

$$sV_C(s) = \frac{1}{C} I_L(s) - \frac{1}{RC} V_C(s) \tag{3.8}$$

Since the transfer function is defined as the ratio of the Laplace Transform of output to that of input, we have to find the ratio of $V_c(s)$ to d(s). From (3.8), we get

$$V_C(s) C\left(s + \frac{1}{RC}\right) = I_L(s) \tag{3.9}$$

Substituting this in (3.7), we will get

$$\frac{v_o(s)}{d(s)} = \frac{\frac{V_i}{LC}}{s^2 + \frac{1}{RC}s + \frac{1}{LC}} \tag{3.10}$$

3.3 Simulation of Buck Converter in Open Loop

The transfer function obtained as (3.10) can be simulated using MATLAB/SIMULINK to obtain the output voltage v_o for various values of duty cycles. The following parameters are used for simulation.

Vi = 100V,

R=1Ω,

L=1mH,

C=1microfarad.

For these values, the transfer function will be

$$\frac{10^{11}}{s^2+10^6 s+10^9}$$

This can be simulated using the transfer function block in SIMULINK. The input duty cycle d can be represented by a constant block and the output voltage can be viewed using a scope block. The SIMULINK diagram is shown in **Fig. 3.4**.

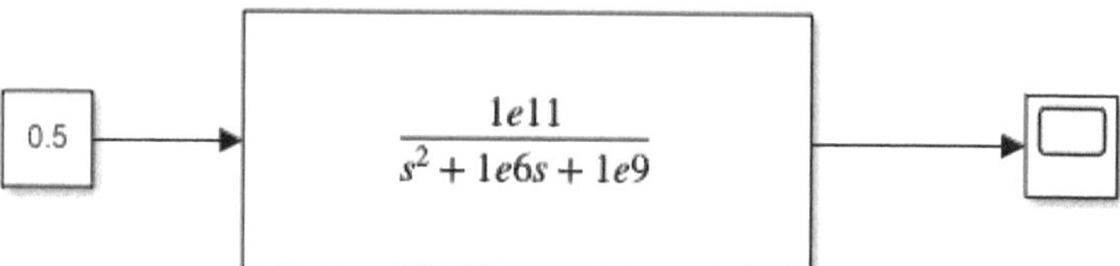

▲ **Fig. 3.4:** SIMULINK diagram of the buck converter

After simulating the system, the output voltage can be observed in the scope block as shown in **Fig. 3.5**. It can be observed from the transient behaviour that the output reaches its final value only after a certain time period.

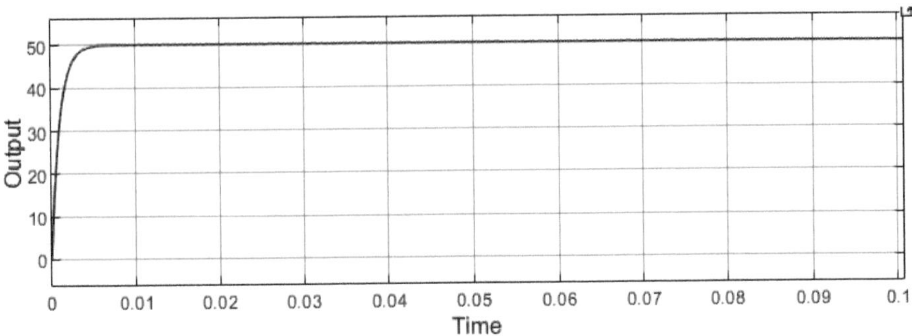

▲ **Fig. 3.5:** Output voltage waveform for d=0.5

3.4 Simulation in Closed Loop

In order to manipulate the transient part of the output response and keep the output constant for any change in the input, closed-loop control is resorted to. In closed loop mode, the duty cycle is continuously adjusted to keep the output constant, by a controller. The simplest controller which gives zero steady-state error is a proportional-integral (PI) controller. By incorporating a PI controller, the simulink diagram gets modified, as shown in **Fig. 3.6**. The desired output voltage is set using the constant block. We have to find the appropriate values of the controller constants by using the PID-tuning facility in MATLAB/SIMULINK.

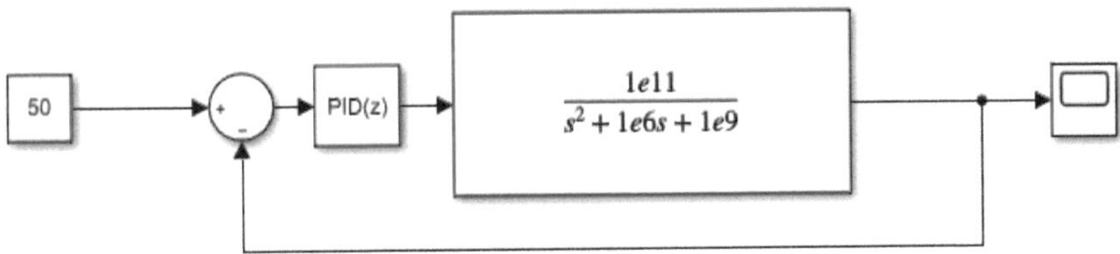

▲ **Fig. 3.6:** SIMULINK diagram for closed-loop control

In the PID block, the default values for P, I and D constants are 1,1 and 0, respectively. The simulation result with these default values is shown in **Fig. 3.7**. Even though the output voltage settles to the reference value, it is observed that there is an overshoot in the beginning. In some cases, overshoots cannot be tolerated. In such cases, proper tuning of the P, I and D values are necessary. In order to tune the PID controller, click open the PID block. The practical implementation of almost all power electronic systems will be in the digital mode with microcontroller-based controllers. Hence, make the PID controller in digital mode by checking the appropriate options in the block. The sample time must be the same as the sampling time with which we would be implementing the system. In our case, the sampling time is taken as 0.01 milliseconds. By clicking the "Tune" button, SIMULINK will come out with a tentative list of controller parameters. These values can be fine-tuned again to get the transient response to our requirement by clicking the "Tuning Tools" menu in the PID tuner. The tuned values obtained here are: P = 0.00721, I = 8.659 and D = 0. The modified response with these parameters is shown in **Fig. 3.8**.

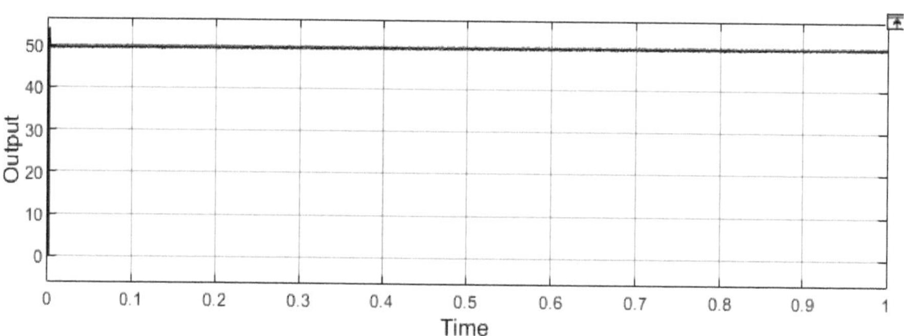

▲ **Fig. 3.7:** Output response with default values for controller constants

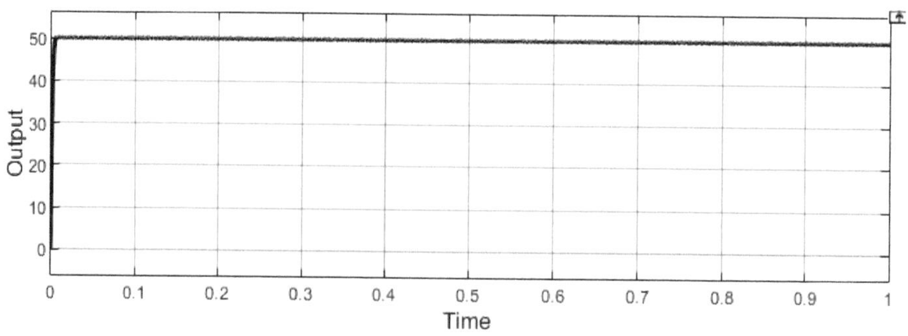

▲ **Fig. 3.8:** Output response with tuned controller parameter values

3.5 Practical Implementation of a Buck Converter

Any power electronic system can be divided into two parts. They are: (i) power circuit and (ii) control circuit. All components which operate at high voltages and currents constitute the power circuit. Semiconductor switches, diodes, inductors and capacitors are part of the power circuit. The circuit for the generation of firing pulses is called the control circuitry. The power circuit is purely analogue in nature and the control circuit is digital. Sensors form the interface between these domains. The high-voltage power circuit and low-voltage control circuit are separated by appropriate isolation circuits like pulse transformers, isolation transformers and optocouplers.

3.5.1 Implementation In The Open Loop

The buck converter consists of only one power electronic switch. In open loop control, the duty of the control circuit is to generate a train of pulses with the desired duty cycle. This can easily be done using any simple microcontroller board. A simple Arduino programme to generate pulses for a 20% duty cycle equivalent to a gain of 0.2 is given below:

Programme 3.1

```
void setup() {
pinMode(3,OUTPUT);
}

void loop() {
digitalWrite(3,HIGH);
delayMicroseconds(200);
digitalWrite(3,LOW);
delayMicroseconds(800);
}
```

A pulse train of 1 kHz with an ON time of 200 microseconds and an OFF period of 800 microseconds will be generated at the 3rd digital pin of the Arduino UNO board. This pulse will be given to the optocoupler input and the output can be given to the gate driver of the IGBT switch. A typical IGBT driver chip is TLP 250, which provides opto isolation as well.

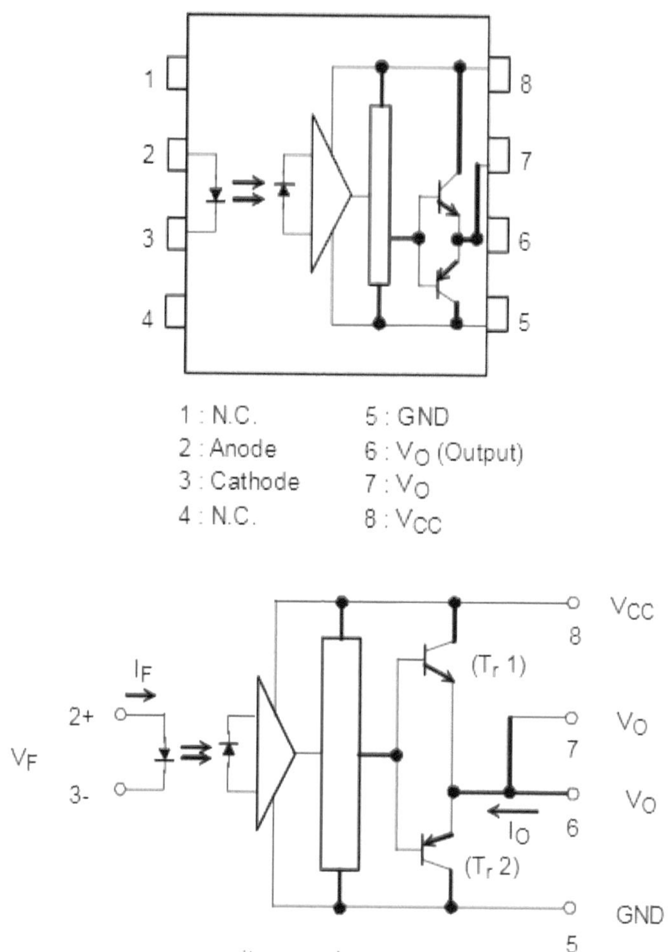

▲ **Fig. 3.7:** Pin configuration and schematic of TLP250 IGBT driver

The Vcc supply voltage will be made using a step-down transformer so that the gate-triggering pulse output from the driver will be isolated. If more than one power electronic switch is present in a circuit, each gate-triggering pulse must be isolated in order to prevent the short-circuiting of the emitter terminals of switches. In advanced driver boards, built-in power supplies exist so that there is no need for external transformers.

The output waveform obtained from the hardware circuit is shown in **Fig. 3.8.**

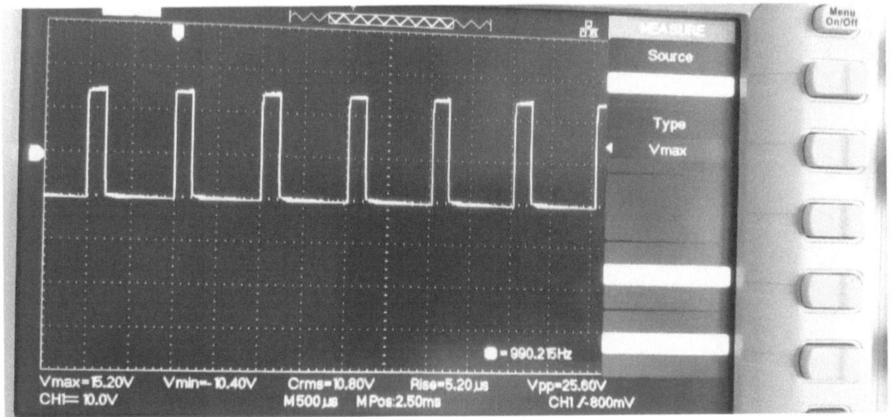

▲ **Fig. 3.8:** Triggering pulses generated with 20% duty cycle

Another programme to generate the same pulses using the PWM feature of Arduino is given below:

Programme 3.2

```
int dutyCycle=25;
void setup() {
 pinMode(3,OUTPUT);


}
void loop() {
 analogWrite(3,dutyCycle);
}
```

The value of the variable duty cycle can take any value from 0 to 255 for varying the duty cycle from 0 to 1. The frequency of the pulses will be 490 Hz at pin 3. In order to have a switching frequency of our choice, we have to install the Arduino PWM library. A programme which can generate PWM signals at any frequency is given below:

Programme 3.3

```
#include <PWM.h>
int32_t frequency = 1000;
int dutyCycle = 50;
void setup() {
 pinMode(3,OUTPUT);
InitTimersSafe();
SetPinFrequencySafe(3, frequency);// set 3rd pin freuency as 1kHz:
 }

void loop() {
  analogWrite(3,dutyCycle);
}
```

If it is required to give the duty cycle as an input in order to vary it continuously, it is possible to read the value of the required duty cycle using a potentiometer connected between the 5 V pin and the GND pin of Arduino. The variable point of the potentiometer is connected to any one of the analogue input pins. The Analogue to Digital Converter within Uno is of 10-bit resolution and the value read from the analogue pin will vary from 0 to 1023 for input voltage varying from 0 to 5 V. This has to be divided by 4 to make the duty cycle within the range of 0 to 256.

Programme 3.4

```
#include <PWM.h>
int32_t frequency = 1000;
int dutyCycle;
void setup() {
 pinMode(3,OUTPUT);
InitTimersSafe();
SetPinFrequencySafe(3, frequency);        // set 3rd pin freuency as 1kHz:
 }

void loop() {
  dutyCycle=analogRead(A0)/4;
  analogWrite(3,dutyCycle);
}
```

3.5.2 Implementation in Closed Loop

The output voltage of any system can be made constant immaterial of any change in input voltage or load by controlling the system in a closed loop. The output is compared with its reference value and the error is passed through a controller. A PI controller is the most commonly used type of control. The output voltage can be sensed either using a hall effect sensor or by making use of a potential divider. If we are using an Arduino Uno board as a controller, we have to scale down the output voltage up to a maximum of 5 V. If it is an Arduino Due board, it has to be limited to 3.3 V. The programme for comparing this voltage with the reference voltage and subjecting the error to a PI controller to find the duty cycle to generate appropriate pulses with the required duty cycle for triggering the switch is given below.

Programme 3.5

```
#include <PWM.h>    // include PWM library:
#include <TimerOne.h>           // include timer library:
int32_t frequency = 1000;    //Set freqency to 1kHz:

float kp = .001;       // PI constants:
 float ki = 0.15;      // PI constants:
// define variables:
float elapsedTime;
float error;
float lastError;
float a,b, out,c, setPoint,d ;
float cumError;
float pivalue;

void setup() {
  pinMode (3, OUTPUT);                              //set 3 pin as output :
InitTimersSafe();
SetPinFrequencySafe(3, frequency);             // set 3rd pin freuency as 1kHz:
   Timer1.initialize(100);
  Timer1.attachInterrupt(computepid);   // function for creating interrupt :
Serial.begin(9600);
}
void loop()
{
}
```

```
void computepid()            // interrupt loop :
 {
  float setPoint = 250.0;     //  set the reference voltage to setpoint:
  float kp = 0.001;           // PI constants:
float ki = 0.15;              // PI constants:
 a = analogRead(A0);          //Read out voltage to A0 (reading analog voltage
 c=(a/4);                   // convert the value to 0 to 255:
 b = ((c*25)/(256))*12.0;     // compute  the actual output voltage
 d = b*1.25;  //Transducer constant
 elapsedTime = .0001;         // Sampling period

              error = setPoint - d;                 // Determine error :
        cumError += error * elapsedTime;      // Compute integral :

      pivalue = kp*error + ki*cumError;       //  PI output:

        lastError = erro;                // remember current error:
        out = (pivalue * 256);                // compute the duty cycle:
analogWrite(3, out);                   // control the output based on PI value:
   }
```

3.6 Boost Converter

The control circuitry and Arduino programmes developed for a buck converter can be used for developing a boost converter, directly, without any change. The only difference here is in the power circuit, where the position of the switch is changed, as shown below in **Fig. 3.9**.

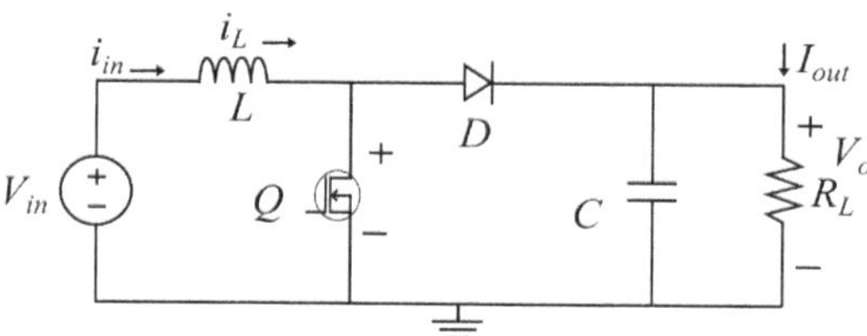

▲ **Fig. 3.9:** Boost Converter

The output voltage V_o from a boost converter will be more than the input voltage V_{in}. As in the case of a buck converter, a boost converter also consists of an active switch Q and a passive switch D. The active switch Q can be made to be ON or OFF at our will with any duty cycle, d. The state of the passive switch will depend on the voltage coming across that switch. When the switch is made ON by applying a positive voltage at its gate, the inductor gets charged. The voltage across the inductor will be V_{in} in this case, as shown in **Fig. 3.10**. When the switch is turned OFF, the stored energy in the inductor will get discharged through the diode. The voltage across the inductor during the turn-off period will be (V_{in}–V_o), as shown in **Fig. 3.11**.

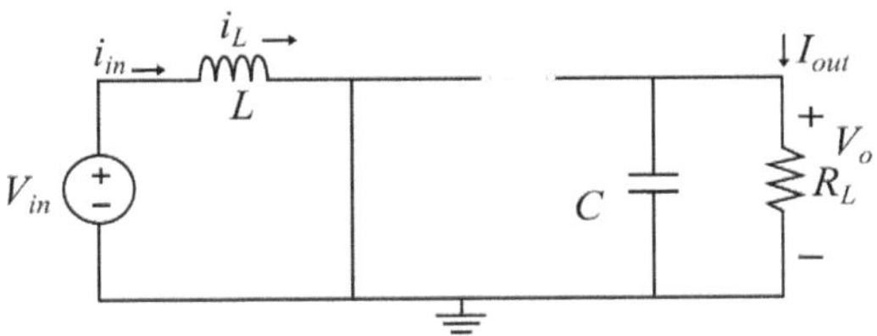

▲ **Fig. 3.10:** Boost converter when the active switch is ON

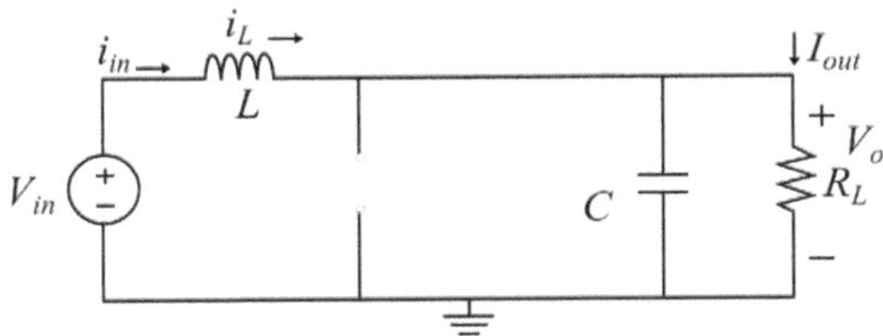

▲ **Fig. 3.11:** Boost converter when the active switch is OFF

Applying the volt-sec balance equation in this case,

$$V_{in}d + (V_{in} - V_o)(1-d) = 0 \tag{3.11}$$

Solving this, the voltage gain can be obtained as

$$\frac{V_0}{V_{in}} = \frac{1}{1-d} \tag{3.12}$$

The programmes 3.1 to 3.4 used for a buck converter can be used for a boost converter as well for open loop operations. The programme for closed loop operation can also be used with a boost converter, but, one has to be a little bit cautious about the range of duty cycle with a boost converter. From equation 3.12, when the duty cycle approaches 1, the denominator will approach zero and the gain will be very large. In order to restrict the output voltage to permissible levels, we have to limit the value of the duty cycle. Hence, programme 3.5 used for closed loop operation of the buck converter is modified in programme 3.6 by incorporating a duty cycle limit.

Programme 3.6

```
#include <PWM.h>   // include PWM library:
#include <TimerOne.h>          // include timer library:
int32_t frequency = 1000;   //Set freqency to 1kHz:
float kp = .001;     // PI constants:
 float ki = 0.15;    // PI constants:

// define variables:
float elapsedTime;
float error;
float lastError;
float a,b, out,c, setPoint,d ;
float cumError;
float pivalue;

void setup() {
  pinMode (3, OUTPUT);                              //set 3 pin as output :
InitTimersSafe();
SetPinFrequencySafe(3, frequency);            // set 3rd pin freuency as 1kHz:

   Timer1.initialize(100);
  Timer1.attachInterrupt(computepid);  // function for creating interrupt :
Serial.begin(9600);
}
void loop()
{
}
```

```
void computepid()          // interrupt loop :
 {
  float setPoint = 250.0;   //  set the reference voltage to setpoint:
  float kp = 0.001;         // PI constants:
float ki = 0.15;            // PI constants:
 a = analogRead(A0);       //Read out voltage to A0 (reading analog voltage
 c=(a/4);                 // convert the value to 0 to 255:
 b = ((c*25)/(256))*12.0;   // compute  the actual output voltage
 d = b*1.25;  //Transducer constant
   elapsedTime = .0001;     // set elapsed time :
       error = setPoint - d;           // Determine error :
       cumError += error * elapsedTime;    // Compute integral :
     pivalue = kp*error + ki*cumError;    //  PI output:
       lastError = error;          // remember current error:
       out = (pivalue * 256);          // compute the duty cycle:

       if (out > 125){ // Limiting the value of duty cycle.
         out = 125;
       }
analogWrite(3, out);                // control the output based on PI value:
}
```

3.7 Buck-Boost Converter

A buck-boost converter gives an output voltage lower than the input voltage for a duty cycle up to 0.5 and an output voltage higher than the input voltage for duty cycles greater than 0.5. The power circuit diagram of a buck-boost converter is shown in **Fig. 3.13**. The direction of the diode is reversed in the case of a buck-boost converter. The diode will conduct when the switch is OFF and in order for this to happen, the polarity at the output side has to be reversed, as shown in the figure. Hence, the polarity of a buck-boost converter with this topology will be opposite to that of the input voltage.

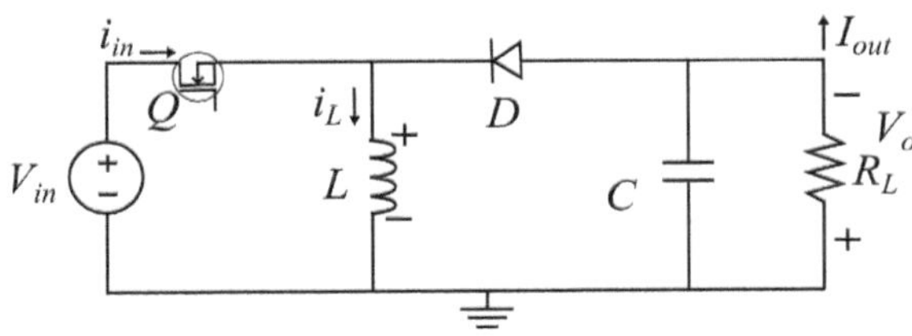

▲ **Fig. 3.12:** Power Circuit of Buck-Boost Converter

The voltage gain equation of a buck-boost converter can be derived in the same manner with which the voltage gain of a buck or boost converter was derived in earlier sections. The volt seconds balance principle is applied here as well. When the switch Q is ON, the inductor gets magnetically charged and the diode will be reverse biased. When the switch Q is turned off, the inductor current will force its way through the diode as shown in the following figures.

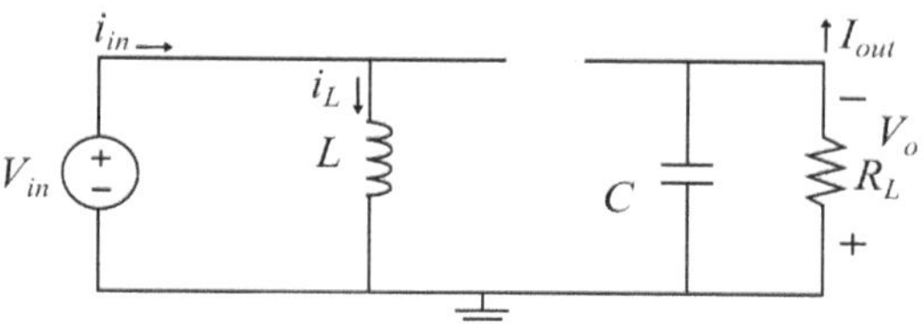

▲ **Fig. 3.13:** Equivalent circuit when Q is ON

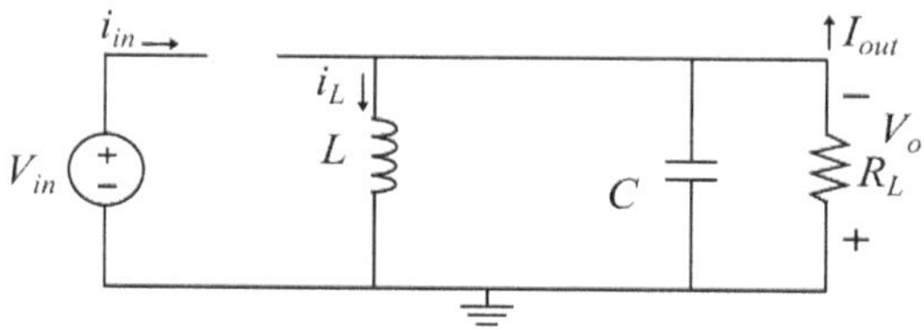

▲ **Fig. 3.14:** Equivalent circuit when Q is OFF

The voltage across the inductor when Q is ON is V_{in} and when Q is OFF, the voltage is $(-V_o)$. Hence, the volt-seconds balance equation becomes

$$V_{in} d - V_o (1-d) = 0 \tag{3.13}$$

Voltage gain can be found as

$$\frac{V_o}{V_{in}} = \frac{d}{(1-d)} \tag{3.14}$$

The same programmes used for buck or boost converter can be used in the case of buck-boost converters as well. But, the upper limit of the duty cycle should be appropriately chosen.

3.8 Design of Inductor

The value of the inductor should be so chosen that the circuit is not driven into discontinuous conduction. This is ensured by choosing an appropriate value for the ripple current. The current through the inductor will be the sum of the DC load current and the ripple current. The ripple current will be passed through the capacitor and only the DC will be passed through the load, ideally. If we assume large values for ripple current, the capacitor value has to be proportionately large to have a pure DC through the load. Hence, a moderate value of 25% ripple current is normally chosen.

In the case of a buck converter, the voltage across the inductor when the switch is OFF is V_o. During this OFF period of (1–d)T, where T is the duration of one cycle, the change in inductor current will be the peak-to-peak value of the ripple current, ΔI. The equation for the voltage across an inductor is

$$V_L = L\frac{di}{dt} \tag{3.15}$$

Substituting,

$$V_o = L\frac{\Delta I}{(1-d)T} \tag{3.16}$$

Hence, the equation for the inductor is

$$L = \frac{V_o(1-d)}{\Delta I\, f} \tag{3.17}$$

where f = 1/T, the switching frequency.

3.9 Design of Capacitor

Similarly, using the current equation of the capacitor, the value of the capacitance required can be f out. The current equation is

$$i_C = C\frac{dV}{dt} \tag{3.18}$$

The current through the capacitor is the ripple current, without any DC component, as shown in **Fig. 3.15.**

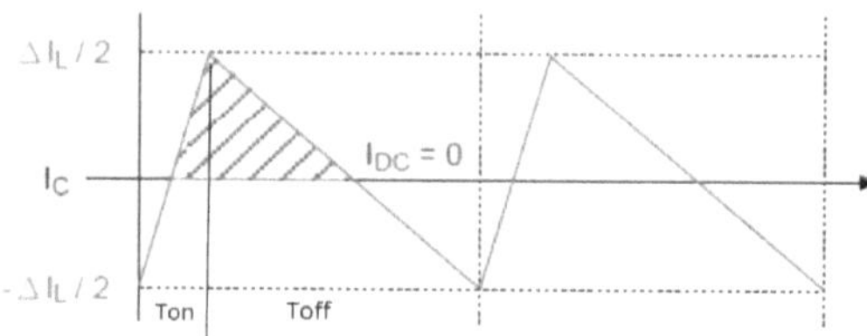

▲ **Fig. 3.15:** Current through the capacitor.

From equation (3.18), we get

$$i_C = C\frac{\Delta V_{CL}}{\Delta T} \tag{3.19}$$

This can be written as

$$i_C \Delta T = C\Delta V_C \tag{3.20}$$

Since the product of current and time is charge,

$$\Delta Q = C\Delta V_C \tag{3.21}$$

During the charging period, the current will be positive. This period is shown shaded in **Fig. 3.15**. The area under the shade will be the charge in one half-cycle. During the other half period, the capacitor discharges. The area of the shaded region will be ΔQ.

$$\Delta Q = \frac{1}{2}\left(\frac{T_{on}}{2} + \frac{T_{off}}{2}\right)\frac{\Delta I_L}{2} = C\,\Delta I_L \tag{3.22}$$

$$C\,\Delta V_C = \frac{1}{2}\left(\frac{DT}{2} + \frac{(1-D)T}{2}\right)\frac{\Delta I_L}{2} \tag{3.23}$$

$$C\,\Delta V_C = \frac{1}{2}\frac{T}{2}\frac{\Delta I_L}{2} \tag{3.24}$$

$$C = \frac{\Delta I_L}{8\,f\,\Delta V_C} \tag{3.25}$$

A similar procedure can be adopted for the design of inductors and capacitors of boost and buck-boost converters.

04 DC-AC CONVERTERS (INVERTERS)

Inverters are DC to AC converter circuits. A fixed DC supply can be converted to fixed/variable voltage and fixed/variable frequency AC voltage. This is achieved by continuously changing the direction of voltage across the load by appropriately switching the power electronic switches ON and OFF. The principle of operation of an inverter can be understood by explaining the working principle of a half-bridge inverter.

4.1 Half-Bridge Square Wave Inverter

The diagram of a half-bridge inverter is shown in **Fig. 4.1**

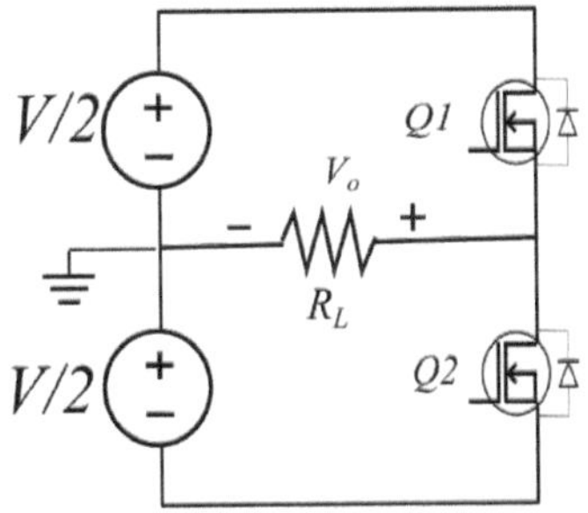

▲ **Fig. 4.1:** Half-Bridge Inverter

The switch Q1 is turned ON and Q2 is turned OFF for a duration equal to half the time period of the AC wave required. Then, the voltage coming across the load R_L will be V/2. For the next half-cycle, Q1 is turned OFF and Q2 is turned ON. In this case, the voltage across the load will be –V/2. If this is repeated, the voltage across the load will be a square wave with an amplitude of V/2, as shown in **Fig. 4.2**. Since the output voltage is a square wave, a half-bridge inverter operated in this manner is called a square wave inverter.

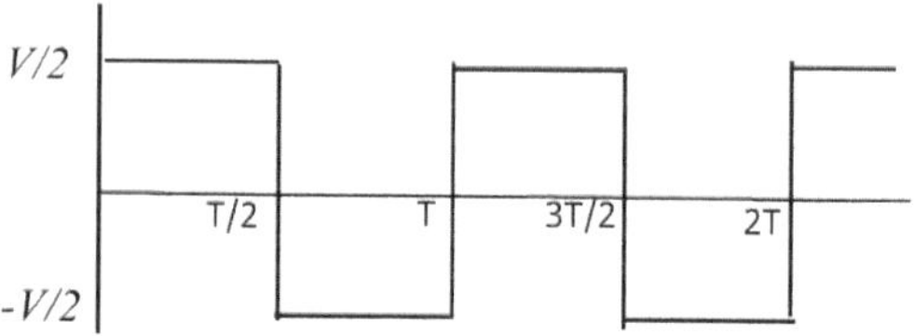

▲ **Fig. 4.2:** Load voltage in a half-bridge inverter

For the operation of a half-bridge inverter, two complementary pulses with a 50% duty cycle are to be created, one for the upper switch and the other for the lower switch, as shown in **Fig. 4.3**.

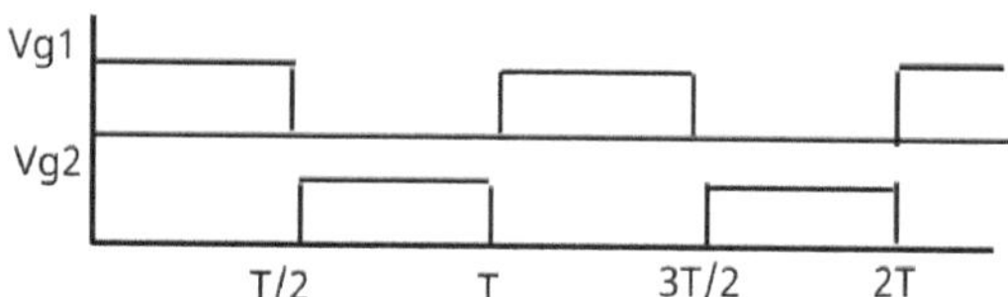

▲ **Fig. 4.3:** Switching pulses for upper and lower switches.

The microcontroller programme is to be written to create one pulse train only, the other pulse train can be generated by passing this pulse train through a NOT gate. However, the generation of the pulses for these two switches in a leg is not that easy due to the finite turn-off period of power electronic switches. Even after making the gate pulse to a switch equal to zero for turn-off, the switch will be actually OFF only after a few nanoseconds or even microseconds of the turn-off period, t_{off}. If we make the lower switch ON at the very moment of making the gate pulse signal equal to zero, both switches will be conducting during the turn-off period. This will create a short-circuit across the DC power supply and the switches will get damaged. This short circuit issue is handled by delaying the turn ON pulse by a short period greater than the turn-off period of the switch. This short period is called the dead band. A dead band can be inserted in the pulse trail either in the software or through a hardware circuit. If the switching frequency is less, it is easier to introduce a dead band in the programme itself. If the switching frequency is large, it is better to do it using a hardware circuit. The hardware circuit for introducing the dead band will be described along with the topic of PWM. After the introduction of the dead band, the pulses applied to the two switches will look like that shown in **Fig. 4.4**.

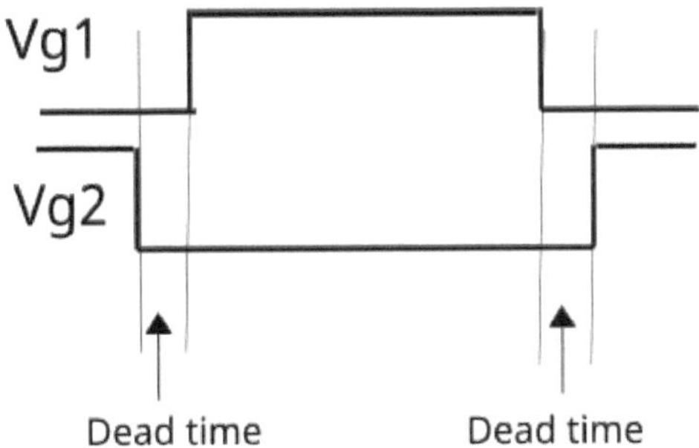

▲ **Fig. 4.4:** Gate pulses with the introduction of dead bands

An Arduino programme to generate gate pulses at a frequency of 50Hz along with the introduction of dead bands is given below:

Programme 4.1

```
void setup() {
  pinMode (3, OUTPUT);
  pinMode (4,OUTPUT);
      }
void loop()
{
digitalWrite(3,HIGH);
digitalWrite(4,LOW);

delayMicroseconds(995); //Duration of first half cycle

digitalWrite(3,LOW);
digitalWrite(4,LOW);

delayMicroseconds(5); //Duration of dead band

digitalWrite(3,LOW);
digitalWrite(4,HIGH);

delayMicroseconds(995); //Duration of second half cycle

digitalWrite(3,LOW);
digitalWrite(4,LOW);

delayMicroseconds(5); //Duration of dead band
}
```

The gate pulses will be available at Digital Pins 3 and 4, which are applied across the gate and emitter terminals of the switches through a gate drive circuit.

4.2 Full-Bridge Square Wave Inverter

A full-bridge inverter will have two legs consisting of an upper switch and a lower switch, as shown in **Fig. 4.5**.

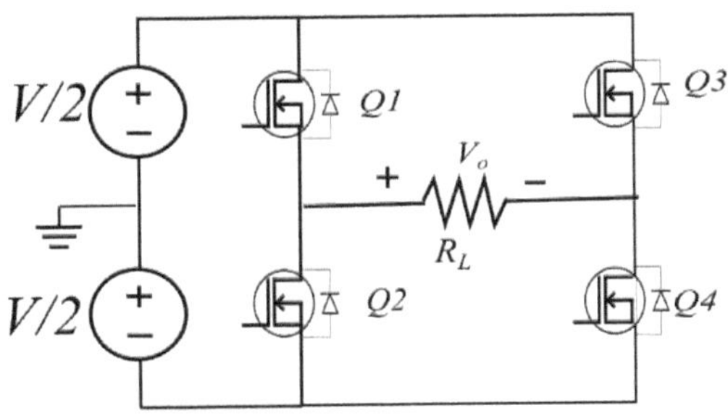

▲ **Fig. 4.5:** Single-phase full-bridge inverter

In this case, for the positive half-cycle, switches Q1 and Q4 are turned ON and Q3 and Q2 are turned OFF. The voltage across the load will be +V with the polarity shown in the figure. For the negative half-cycle, Q3 and Q2 are turned ON and Q1 and Q4 are turned OFF. The voltage across the load will be −V with the polarity shown. The output voltage and gate pulses are shown in **Fig. 4.6**.

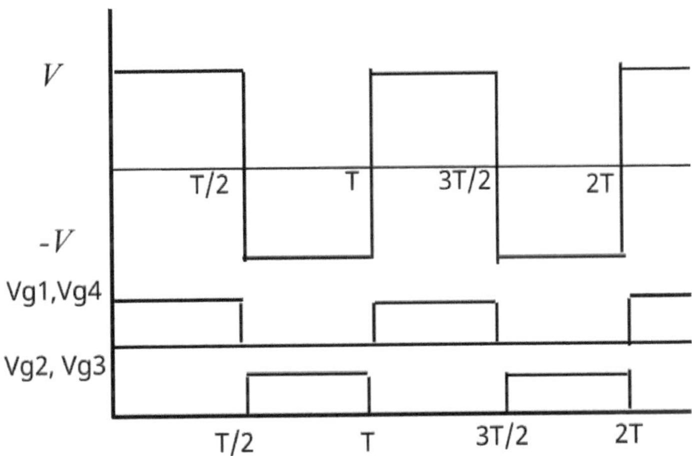

▲ **Fig. 4.6:** Output voltage and gate pulses for full-bridge square wave inverter

The same programme for the generation of pulses in the case of a half-bridge can also be used in the case of a full-bridge inverter. The pulses for Q1 generated for the half-wave inverter can be given to both Q1 and Q4, in this case. Similarly, the pulses for Q2 in the case of a half-bridge inverter can be used for Q3 and Q2 in the case of a full-bridge inverter. Another Arduino programme for the generation of four pulses separately for the four switches including dead time is shown below.

Programme 4.2

```
void setup() {
  pinMode (3, OUTPUT);    //set 3 pin as output :
  pinMode (4,OUTPUT);
  pinMode(5, OUTPUT);
  pinMode(6,OUTPUT);
}
void loop()
{
digitalWrite(3,HIGH);//positive half cycle
digitalWrite(4,LOW);
digitalWrite(5,LOW);
digitalWrite(6,HIGH);
delayMicroseconds(445);
digitalWrite(3,LOW); //deadband
digitalWrite(4,LOW);
digitalWrite(5,LOW);
digitalWrite(6,LOW);
delayMicroseconds(5);
digitalWrite(3,LOW); //negative half cycle
digitalWrite(4,HIGH);
digitalWrite(5,HIGH);
digitalWrite(6,LOW);
delayMicroseconds(445);
digitalWrite(3,LOW); //deadband
digitalWrite(4,LOW);
digitalWrite(5,LOW);
digitalWrite(6,LOW);
delayMicroseconds(5);
}
```

The output voltage obtained from a full-bridge inverter obtained using the above programme is shown in **Fig. 4.7**.

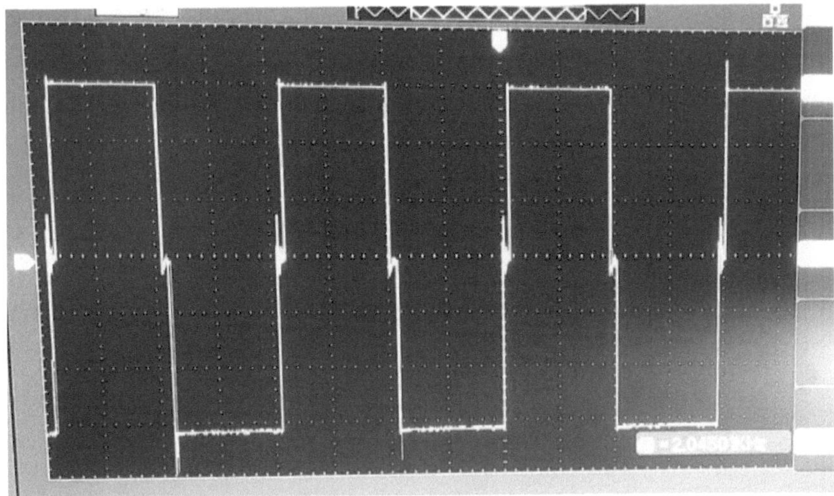

▲ **Fig. 4.7:** Load voltage of a full bridge inverter

For resistive loads, the current waveform will be the same as that of voltage. For inductive loads, the current will be exponentially rising and falling. Depending on the time constant of load, the wave shape will vary. For large values of time constant, one would get a triangular waveform for current. A typical current waveform obtained is shown in **Fig. 4.8.**

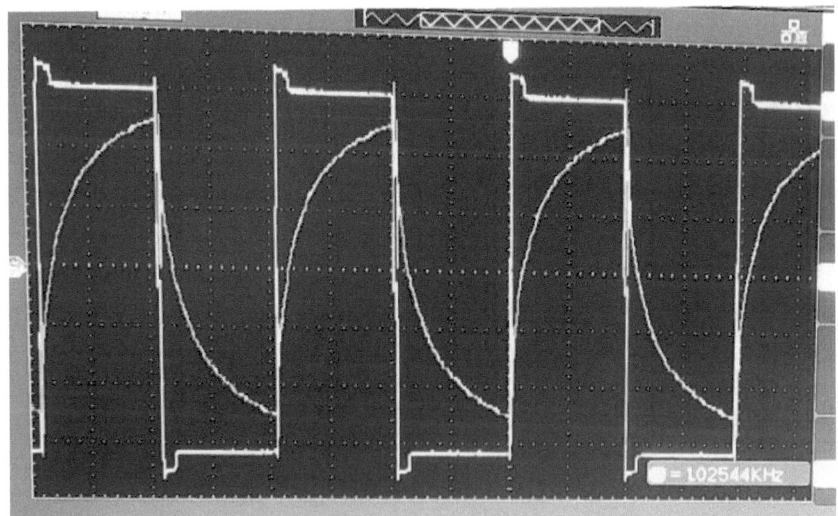

▲ **Fig. 4.8:** Square voltage waveform and exponential current waveform from a full-bridge inverter

When the frequency operation is very large, the dead bands may have to be made less than 1 microsecond, in some cases. It is difficult to provide delays lesser than 1 microsecond using simple programming with Arduino. With some DSPs, there is a provision to include dead bands in the PWM signal pins. One can create dead bands of any duration using simple hardware circuits. One such circuit is given in **Fig. 4.7**.

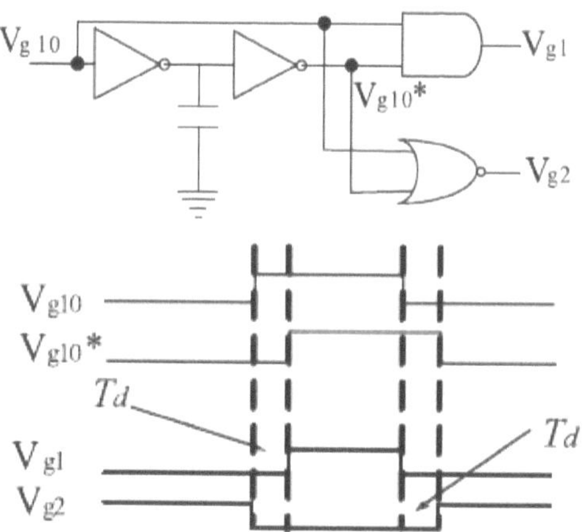

▲ **Fig. 4.7:** Dead band generation circuit and input/output waveforms

If a hardware circuit is used for the dead-time generation, the microcontroller needs to generate only one pulse train for a leg in the inverter. The hardware circuit will generate two pulse trains, one for the upper switch and the other one for the lower switch with the insertion of a dead band in between switching. Here, V_{g0} is the pulse generated from the controller. V_{g1} and V_{g2} are the signals generated from the hardware circuit for the upper and lower switch respectively. T_d is the delay time. One can choose the appropriate value for the capacitor value, which decides the relay time. The waveforms generated for a value of 10 nanofarads are shown in **Fig. 4.8**.

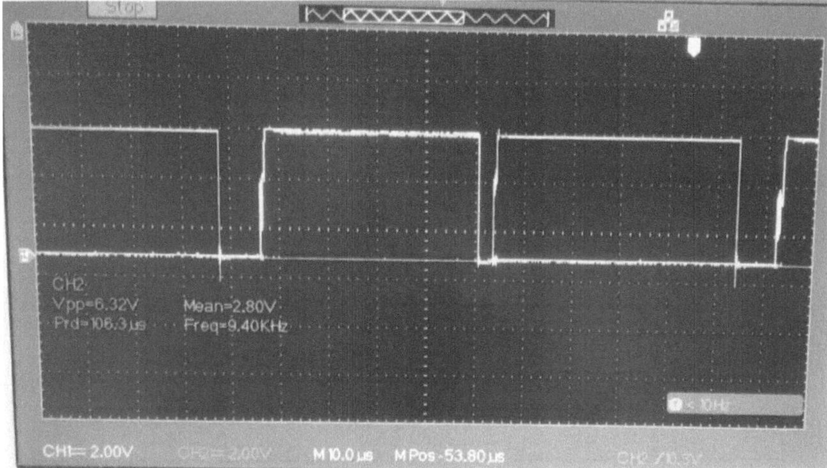

▲ **Fig. 4.8:** Practical wave generated with a hardware dead band circuit

4.3 Three-Phase Inverters

A three-phase inverter can be made by adding one more leg to a single-phase bridge inverter, as shown in **Fig. 4.9**.

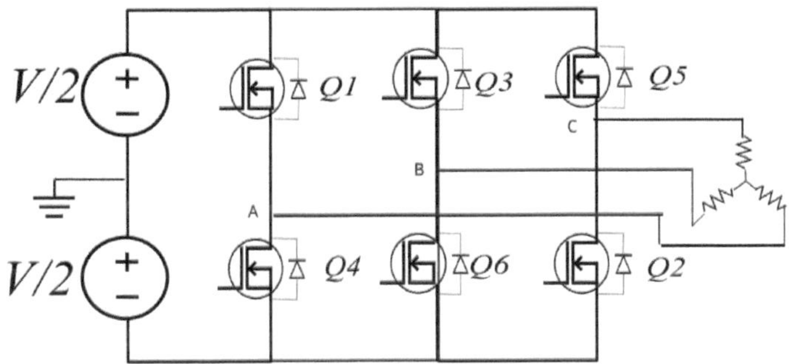

▲ **Fig. 4.9:** Three-phase inverter

There are two modes in which a three-phase inverter can operate. They are (i) 120 degrees conduction mode and (ii) 180 degrees conduction mode. As the name of these modes suggests, every switch will be conducting for a duration of 120° in the cycle in 120° conduction mode. In 180° conduction mode, the duration for conduction of every switch will be equivalent to 180°.

1) 120° conduction mode:

In this mode, gate pulses of 120° duration are to be given to every switch. The pulses for Q1, Q3 and Q5 will be displaced by 120°. Similarly, the gate pulses to Q2, Q4 and Q6 will be phase shifted by 120°. Gate pulses in this case are depicted in **Fig. 4.10**.

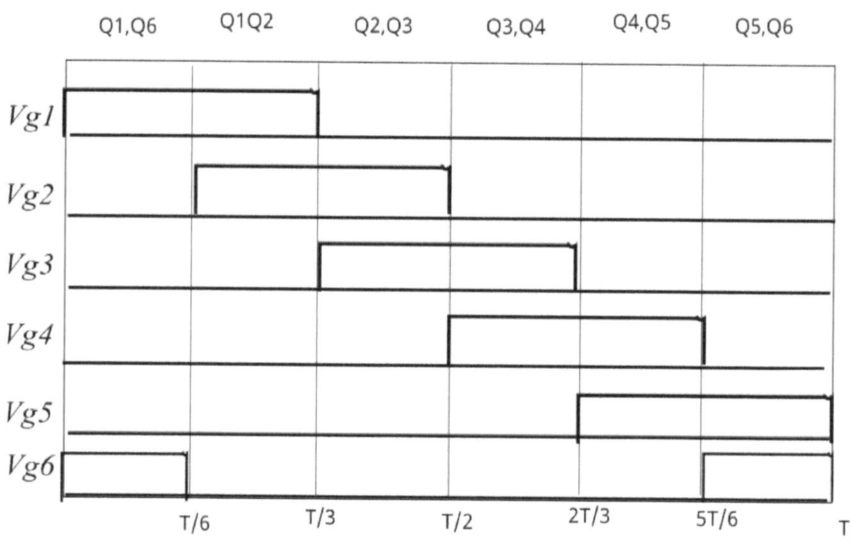

▲ **Fig. 4.10:** Gate pulses in a three-phase square wave inverter

It is evident from the above diagram that a transition of switching state happens in every 60° duration. In every segment of 60° duration, two switches, one on the upper side and one on the lower side, conduct. Moreover, there is a definite gap of 60° between the turn-off of every upper switch and the turn-on of its corresponding lower switch. Hence, the chances of both switches conducting together in a leg leading to the short-circuit of the DC source is not there in the case of a three-phase square wave inverter and there is no need for a dead band generation circuit here. A programme for the generation of switching pulses in the case of a three-phase inverter is given below. PINs 1 to 6 are assigned for triggering switches 1 to 6 respectively.

Programme 4.3

```
void setup() {
  pinMode(1,OUTPUT);
  pinMode(2,OUTPUT);
  pinMode (3, OUTPUT);
  pinMode (4,OUTPUT);
  pinMode(5, OUTPUT);
  pinMode(6,OUTPUT);

digitalWrite(2,LOW);
digitalWrite(3,LOW);
digitalWrite(4,LOW);
digitalWrite(5,LOW);
}
void loop()
{
digitalWrite(1,HIGH);
digitalWrite(6,HIGH);

delayMicroseconds(3333);

digitalWrite(2,HIGH);
digitalWrite(6,LOW);
 delayMicroseconds(3332);

digitalWrite(1,LOW);
digitalWrite(3,HIGH);

delayMicroseconds(3332);

digitalWrite(2,LOW);
digitalWrite(4,HIGH);

digitalWrite(3,LOW);
digitalWrite(5,HIGH);

delayMicroseconds(3332);

digitalWrite(4,LOW);
digitalWrite(6,HIGH);

delayMicroseconds(3332);
}
```

For generating an output voltage having a frequency of 50 Hz, the time period, T is to be 20 milliseconds. Hence, a time period of T/6 will be 3.333 milliseconds or 3333 microseconds.

The midpoint of the supply voltage is called the virtual ground. All voltages will be mentioned with respect to this ground point. The voltages at points A, B and C measured with respect to this virtual ground point are called the pole voltages. The pole voltages will be looking like those shown in **Fig. 4.11**.

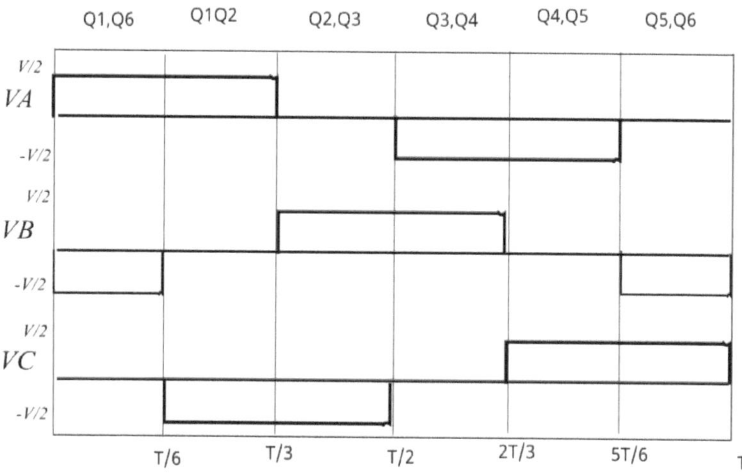

▲ **Fig. 4.11:** Pole voltage waveforms in 120° conduction mode

The line-to-line voltages V_{AB}, V_{BC} and V_{CA} can be found as (V_A–V_B), (V_B–V_C) and (V_C–V_A), respectively, as shown in **Fig. 4.12**.

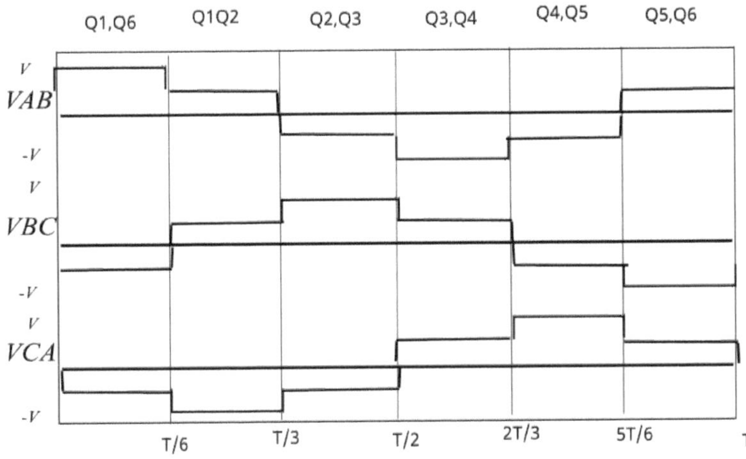

▲ **Fig. 4.12:** Line-to-line voltages in 120° conduction mode

2) 180° conduction mode:

In this mode, every switch will be made conducting for a duration of 180° in a cycle. The firing pulses to be given to the switches are depicted in **Fig. 4.13**. It can be observed that one switch on the upper side and two switches on the lower side or two switches on the upper side and one switch on the lower side conduct at any instant.

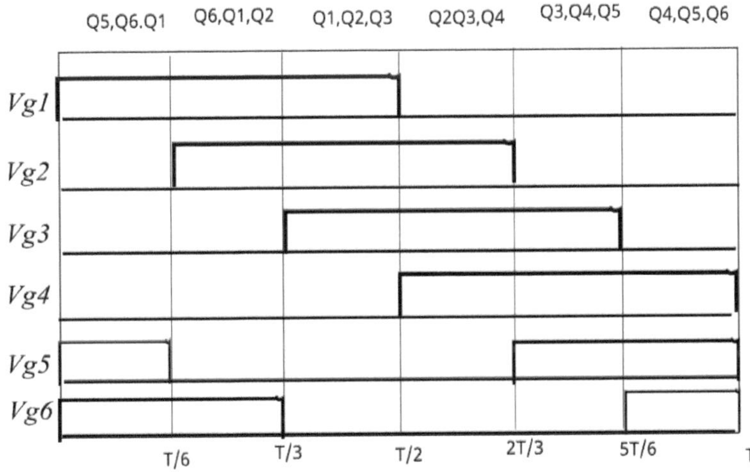

▲ **Fig. 4.13:** Gating pulsed for 180° conduction mode

An Arduino programme to generate these pulses at digital pins from 1 to 6 can be written as

```
void setup() {
  pinMode(1,OUTPUT);
  pinMode(2,OUTPUT);
  pinMode (3, OUTPUT);
  pinMode (4,OUTPUT);
  pinMode(5, OUTPUT);
  pinMode(6,OUTPUT);

digitalWrite(1,LOW);
digitalWrite(2,LOW);
digitalWrite(3,LOW);
digitalWrite(4,LOW);
digitalWrite(5,LOW);
digitalWrite(6,LOW);
}
void loop()
{
digitalWrite(1,HIGH);
digitalWrite(5,HIGH);
digitalWrite(6,HIGH);

delayMicroseconds(3330);

digitalWrite(5,LOW);
delayMicroseconds(3);
digitalWrite(2,HIGH);

delayMicroseconds(3330);
```

```
digitalWrite(6,LOW);
delayMicroseconds(3);
digitalWrite(3,HIGH);

delayMicroseconds(3330);

digitalWrite(1,LOW);
delayMicroseconds(3);
digitalWrite(4,HIGH);

delayMicroseconds(3330);

digitalWrite(2,LOW);
delayMicroseconds(3);
digitalWrite(5,HIGH);

delayMicroseconds(3330);

digitalWrite(3,LOW);
delayMicroseconds(3);
digitalWrite(6,HIGH);

delayMicroseconds(3330);

digitalWrite(4,LOW);
delayMicroseconds(3);
digitalWrite(1,HIGH);

delayMicroseconds(3330);

}
```

Since the simultaneous change in the conduction mode of both switches in the same leg happens here, it is necessary to introduce a dead band between these transitions. A dead time of 3 microseconds is provided in this programme for the legs in which simultaneous transitions happen.

The pole voltages in this case will be as shown in **Fig. 4.14**.

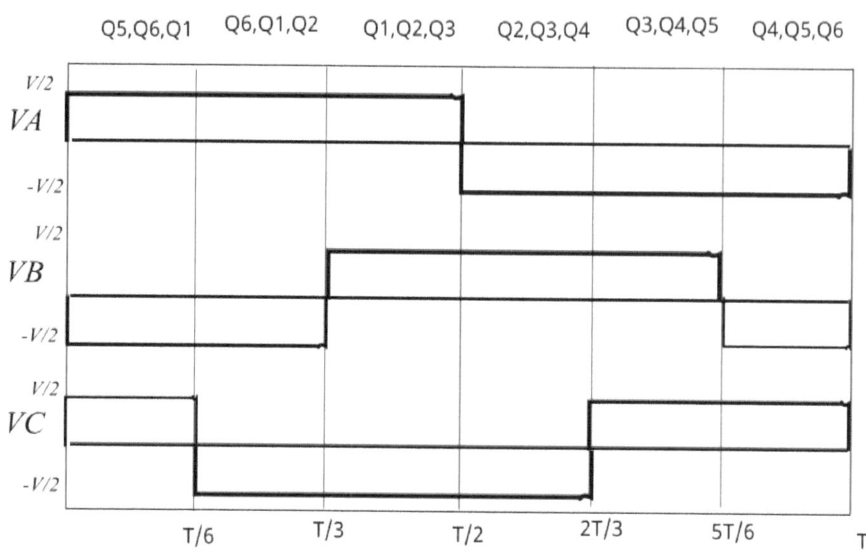

▲ **Fig. 4.14:** Pole voltages in 180° conduction mode

The line-to-line voltages will be evaluated as the difference between the corresponding pole voltages, as shown in **Fig. 4.15**.

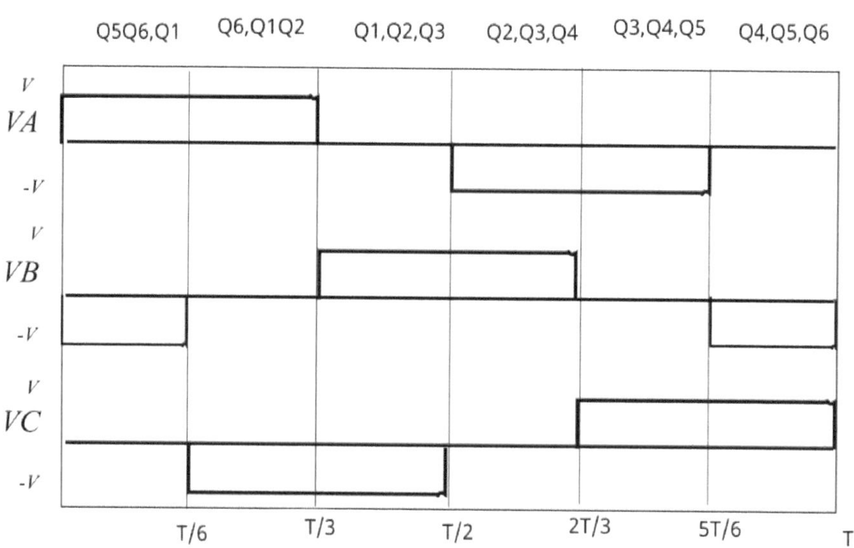

▲ **Fig. 4.15:** Line voltages in 180° conduction mode

4.4 PWM Inverters

Even though the output voltages obtained by the inverter operations described in the above sections are AC in nature, the wave shape is that of a square wave. Ideally, the wave shape expected is a sine wave. Any waveform can be considered as a sum of several sine waves. A square wave can also be split into several sine waves using Fourier Transformation.

A unity square wave can be represented as shown in **Fig. 4.16**.

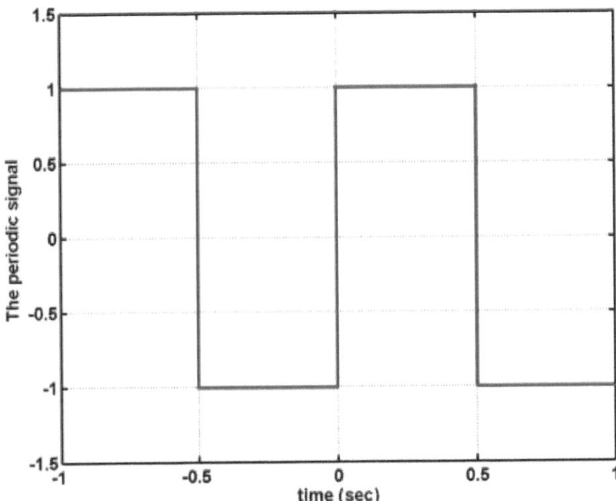

▲ **Fig. 4.16:** Unity square wave

The Fourier Series of this waveform can be written as

$$x(t) = \sum_{n=1,3,5\ldots}^{\infty} \frac{4}{n\pi} \sin(n\omega_o t) \quad (4.1)$$

Putting n=1, the fundamental sine wave will be obtained as 1.27 $\sin(\omega_o t)$, which is plotted along with the square wave, as shown in **Fig. 4.17**.

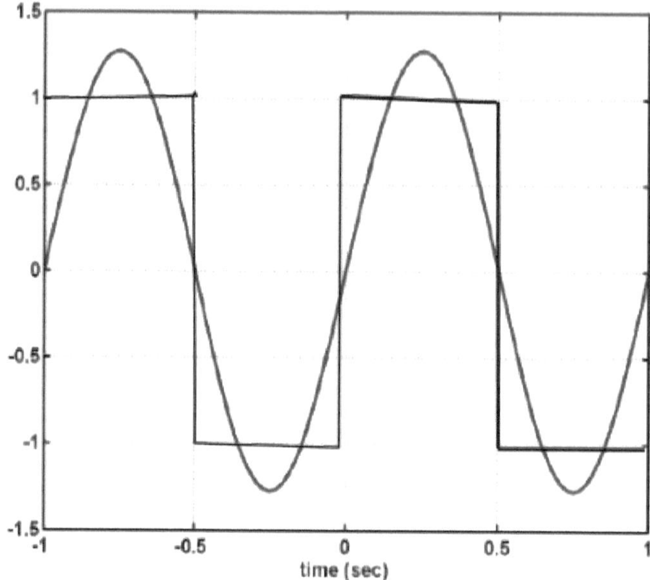

▲ **Fig. 4.17:** Fundamental sine wave

It may be noted that the peak value of the fundamental wave is 1.27, which is higher than the DC value of 1 for the square wave. For odd values of n ranging from 3 to infinity, the peak values for the 3rd, 5th .... Harmonic components can be evaluated. The magnitude of harmonics along with the magnitude of the fundamental can be depicted as shown in **Fig. 4.18.**

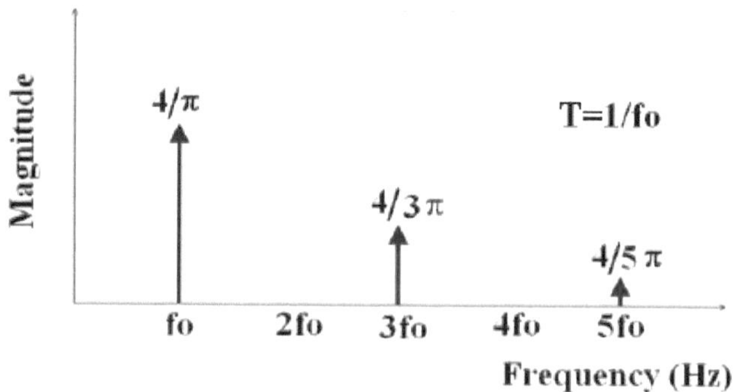

▲ **Fig. 4.18:** Harmonic spectrum of the square wave

All frequency components except for the fundamental are useless as far as the active power is concerned. Harmonic powers are reactive in nature and hence needed to be eliminated by connecting filters. The size of filters required will be inversely proportional to the frequency of components to be removed. For a square wave inverter working at a fundamental frequency of 50 Hz, the harmonics to be filtered out will be of frequencies 150 Hz, 250 Hz etc. These are relatively small frequencies and the size of filter components will be very huge. One way to reduce the filter requirements is by modifying the waveform and eliminating lower-order harmonics like 3rd, 5th etc. This can be achieved by adopting a switching strategy called PWM. In PWM, the switching frequency is increased to a high value. The harmonic components will be only the multiples of the switching frequency. In a square wave, the switching frequency is the same as the fundamental frequency of 50 Hz. In PWM, the switching frequency will be increased to several kilohertz. Hence, the harmonic components will only be in the range of kilohertz, which can be filtered out easily. A small inductive load will be capable of eliminating harmonic currents of this order. A fundamental component of 50 Hz is obtained by changing the duty cycle of switching in a sinusoidal manner with the fundamental frequency. Change in duty cycle in a sinusoidal manner can be easily achieved in hardware by generating two signals. One is a sinusoidal wave at the fundamental frequency, which can be generated using an oscillator circuit. This wave is called the modulation signal. The second signal is a triangular wave operating at a very high frequency, which is called a carrier wave. These two signals will be compared using an op-amp comparator so that the output will be at logical HIGH when the modulation signal is higher than the carrier signal and at logical LOW when the modulation signal is lower than the carrier signal, as shown in **Fig. 4.19**.

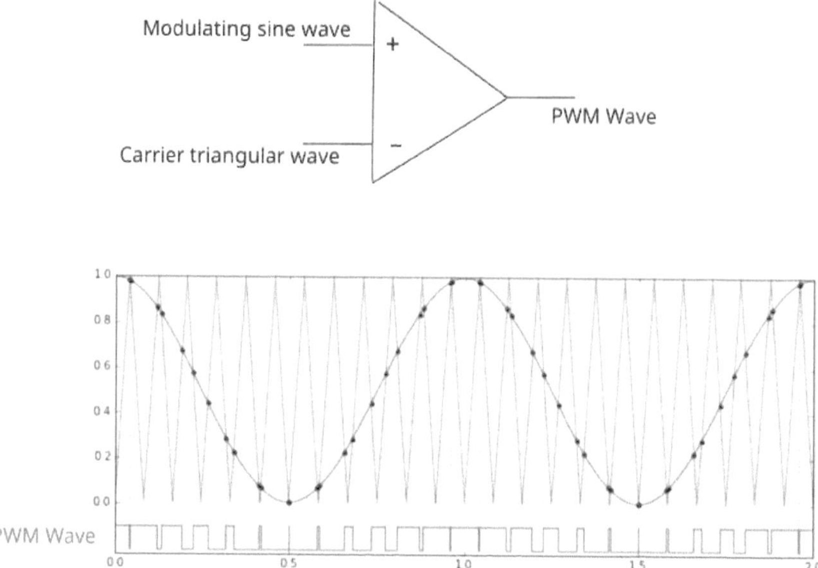

▲ **Fig. 4.19:** PWM wave generation

4.4.1 PWM-Based Half-Bridge Inverter

In a half-bridge PWM inverter, the switching pulses to the upper switch will be the same as the PWM signal. The lower switch is provided with the inverted PWM signal. In order to avoid short-circuit due to the simultaneous conduction of both switches in the same leg, the switching signals are generated using a dead time generator circuit mentioned in **Fig. 6.7**. An Arduino programme for PWM generation is shown below. Rather than creating the PWM signals by comparison of carrier and modulating wave, the duty ratio of the PWM pin is continuously varied by the modulating signal.

```
#include <PWM.h>    // include PWM library:
int32_t frequency = 1000;    //Set freqency

// define variables:
float omega=314.0;
float t;
float out;

void setup() {
  pinMode (3, OUTPUT);                          //set 3 pin as output :
 SetPinFrequencySafe(3, frequency);   // set 3rd pin freuency as 1kHz:
}
void loop()
{
  t=millis();
out=sin(omega*t*0.001);
out=out*125+125;
analogWrite(3,out);
}
```

The gate pulses for the upper and lower switches are depicted in **Fig. 4.20**.

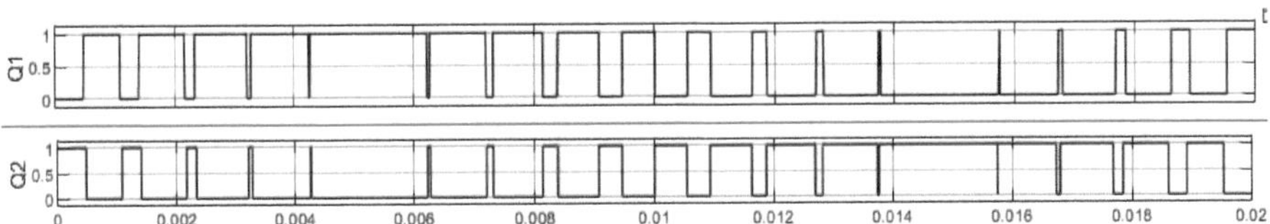

▲ **Fig. 4.20:** Gate pulses for PWM half-bridge inverter

The output voltage from a half-bridge PWM inverter with two voltage sources of 100 V with a midpoint connected to the load will be as shown in **Fig. 4.21**.

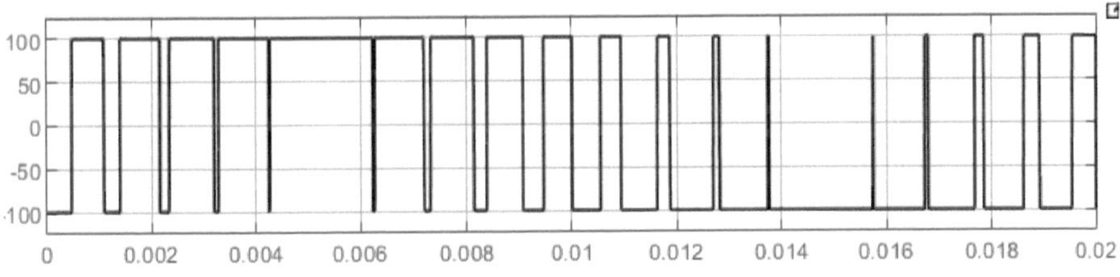

▲ **Fig. 4.21:** Load voltage from a PWM half-bridge inverter

4.4.2 PWM for Full-Bridge Inverters

A single-phase full-bridge inverter given in a previous section is reproduced here in **Fig. 4.22** to explain the working of a PWM full-bridge inverter.

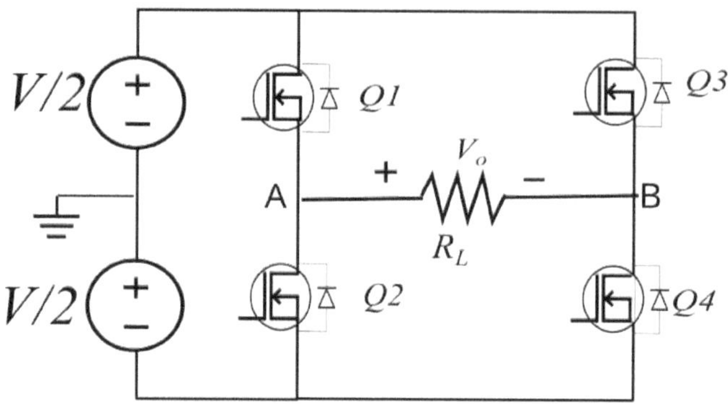

▲ **Fig. 4.22:** Single-phase full-bridge inverter

There are two modes of operation for this full-bridge inverter, namely, bipolar mode and unipolar mode. In bipolar mode, the output voltage V_{AB} will be changing from +V to −V with the switching frequency. But, in unipolar mode, this output voltage will be changing from +V to 0 in the positive half-cycle and from −V to 0 in the negative half-cycle, as shown in **Fig. 4.23.**

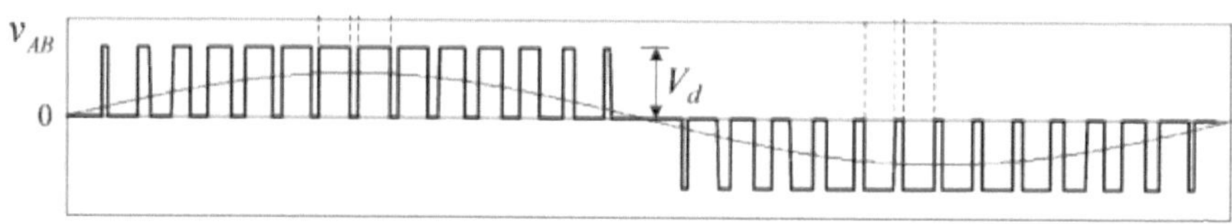

▲ **Fig. 4.23(a):** Output voltage in unipolar operation

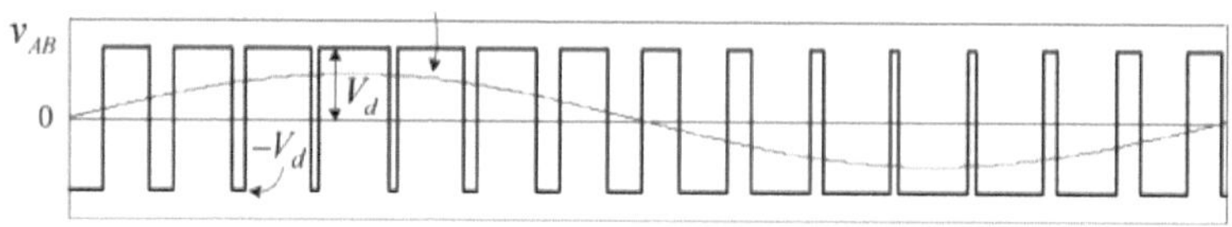

▲ **Fig. 4.23(b):** Output voltage in bipolar operation

4.4.2.1 Bipolar Operation

In bridge inverters, one has to generate two gating pulse trains for the upper switches of each leg. The gating pulses for the lower switches will be the complement of the pulses for the upper switch, with the appropriate dead band in between. In bipolar operation, there will be only one sinusoidal modulating signal. This will be compared with the triangular carrier wave and PWM signals are generated for the upper switch in one leg. The switching pulses for the second leg will be generated by inverting the gate pulses of the first leg. Hence, switches Q1 and Q4 will have the same gate pulses and switches Q2 and Q3 will have the same pulses, which are generated by the inversion of pulses for Q1 and Q4. Hence, the microcontroller needs to generate only one signal. The same programme used in the case of the half-bridge inverter can be used in this case. The gating signals, pole voltages V_A and V_B, and the load voltage V_{AB} are depicted in **Fig. 4.24**.

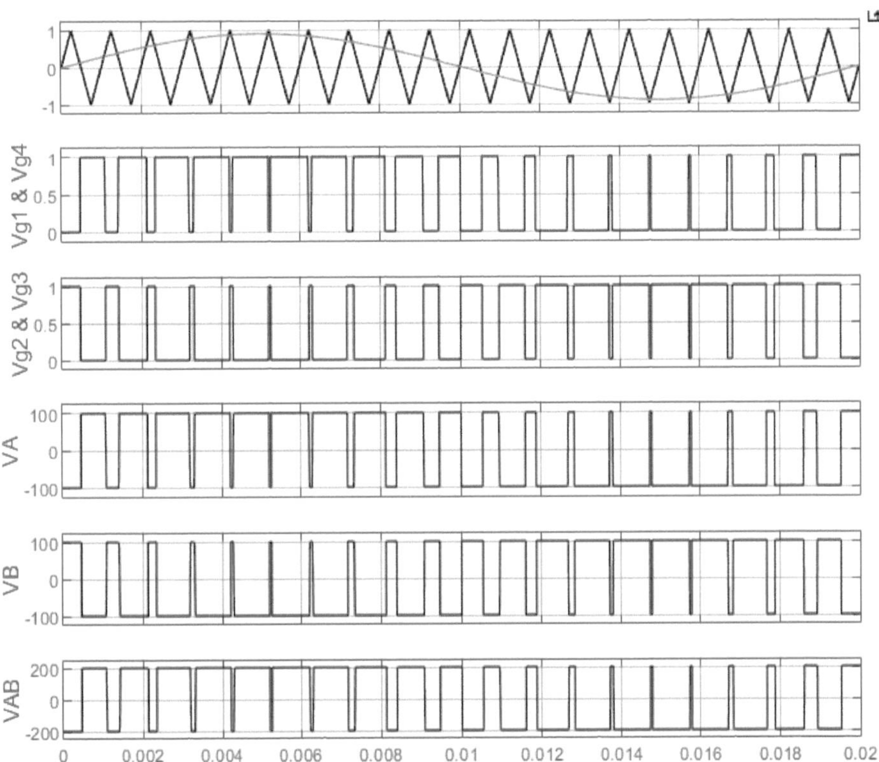

▲ **Fig. 4.24:** Modulating and carrier waves, gating signals, pole voltages and load voltage in bipolar mode

The same programme written for a half-bridge sine PWM inverter can be used for the generation of pulses for bipolar sine PWM operation of a full-bridge inverter. The programme is reproduced below:

```
#include <PWM.h>    // include PWM library:
int32_t frequency = 1000;    //Set freqency

// define variables:
float omega=314.0;
float t;
float out;

void setup() {
  pinMode (3, OUTPUT);                    //set 3 pin as output :
 SetPinFrequencySafe(3, frequency);  // set 3rd pin freuency as 1kHz:
}
void loop()
{
  t=millis();
out=sin(omega*t*0.001);
out=out*125+125;
analogWrite(3,out);
}
```

PWM pulses generated at pin 3 are shown in **Fig. 4.25**.

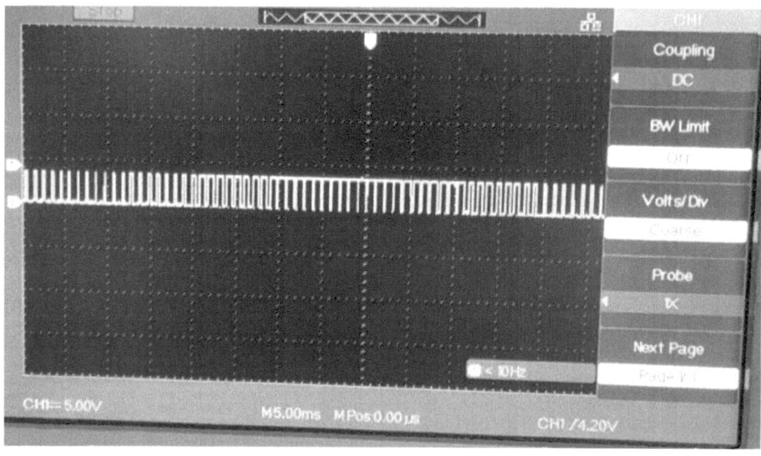

▲ **Fig. 4.25:** Gate pulses generated at pin 3 for sine PWM

Switches T1 and T4 are provided with these pulses for bipolar operation. This signal is passed through a NOT gate and applied at the switches T2 and T3. The output voltage will vary between +V and −V at the load terminals. The waveform obtained is shown in **Fig. 4.26**.

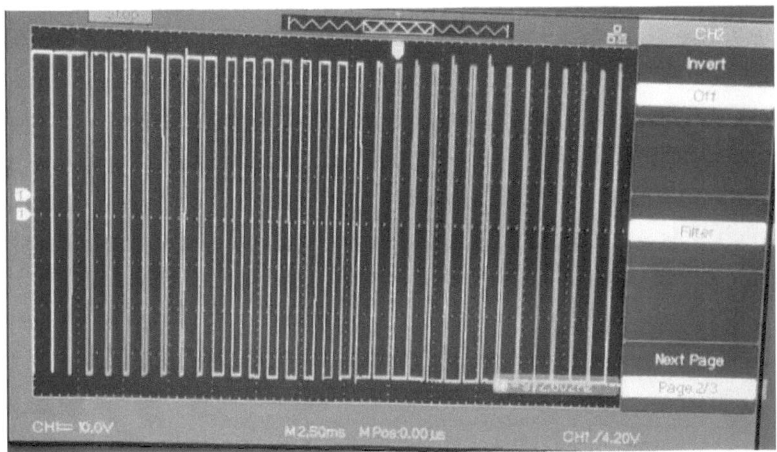

▲ **Fig. 4.26:** Output voltage waveform of a bipolar PWM inverter

If the load is resistive, the current wave will have the same shape as that of the voltage wave. If the load is inductive, the current tends to be sinusoidal, depending on the time constant of the load. The waveform obtained for a low time constant load is shown in **Fig. 4.27**.

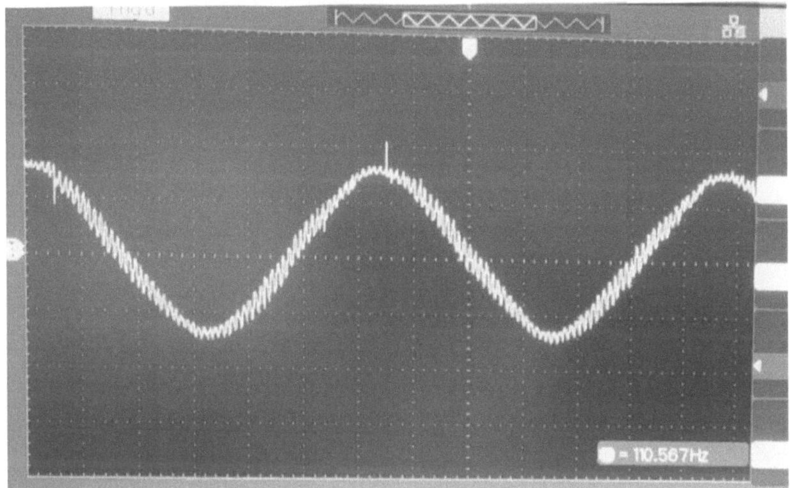

▲ **Fig. 4.27:** Current waveform for an inductive load

It can be observed that when the modulating sine wave is having a value near zero, the duty cycle of the output wave is 0.5 so that the duration for which the output is having a value of +V is the same as the duration for which the output is –V. When the value of the modulating sine wave is positive, the duration for which the output voltage is +V will be more than the duration for which it is –V. The duration for +V reaches a maximum at the peak value of the modulating wave and then it starts decreasing. It becomes equal to the duration of –V again at the zero crossing point of the modulating wave. In the negative cycle, the duration for which the output is –V will be more than the duration for which it is +V. The voltage and current waves obtained are shown together in **Fig. 4.28**.

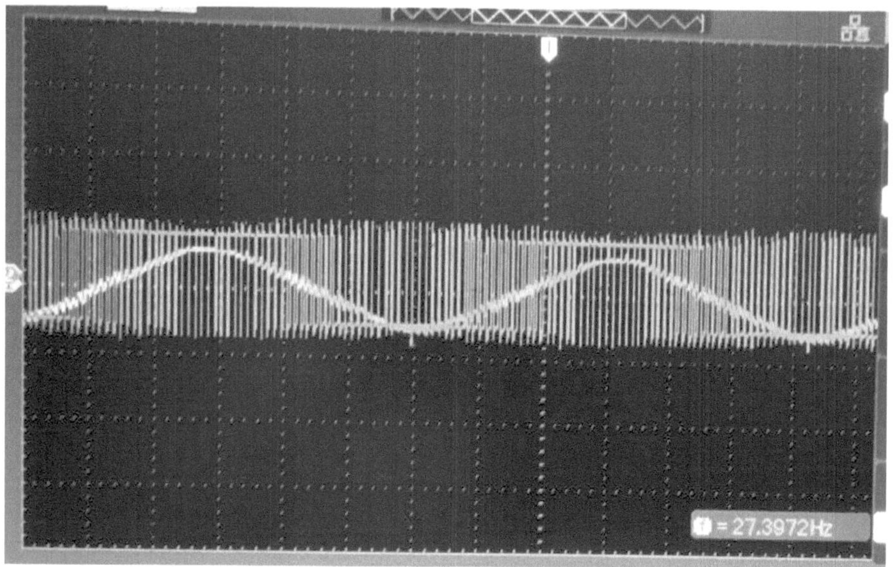

▲ **Fig. 4.28:** Current and voltage waveforms from a sine PWM inverter

4.4.2.2 Unipolar Mode of Operation.

In the unipolar mode of operation, both legs are operated independently. Hence, there will be two modulation signals, one for each leg. The modulation signal for the second leg will be displaced by 180° with respect to that of the first leg. These modulated signals will be compared with a common carrier signal, and two gate signals for the upper switches of both legs are generated. The switching pulses for the lower switches can be generated through dead band circuits. An Arduino programme for the generation of these two pulses is given below:

```
#include <PWM.h>    // include PWM library:
int32_t frequency = 1000;    //Set freqency

// define variables:
float omega=314.0;
float t;
float out1,out2;

void setup() {
  pinMode (3, OUTPUT);                    //set 3 pin as output :
  pinMode (4,OUTPUT);
 SetPinFrequencySafe(3, frequency);   // set 3rd pin freuency as 1kHz:
}
void loop()
{
  t=millis();
out1=sin(omega*t*0.001);
out2=-1*out1;
out1=out1*125+125;
out2=out2*125+125;
analogWrite(3,out1);
analogWrite(4,out2);
}
```

The modulating signals, carrier wave, gating signals, pole voltages and load voltage are depicted in **Fig. 4.29**.

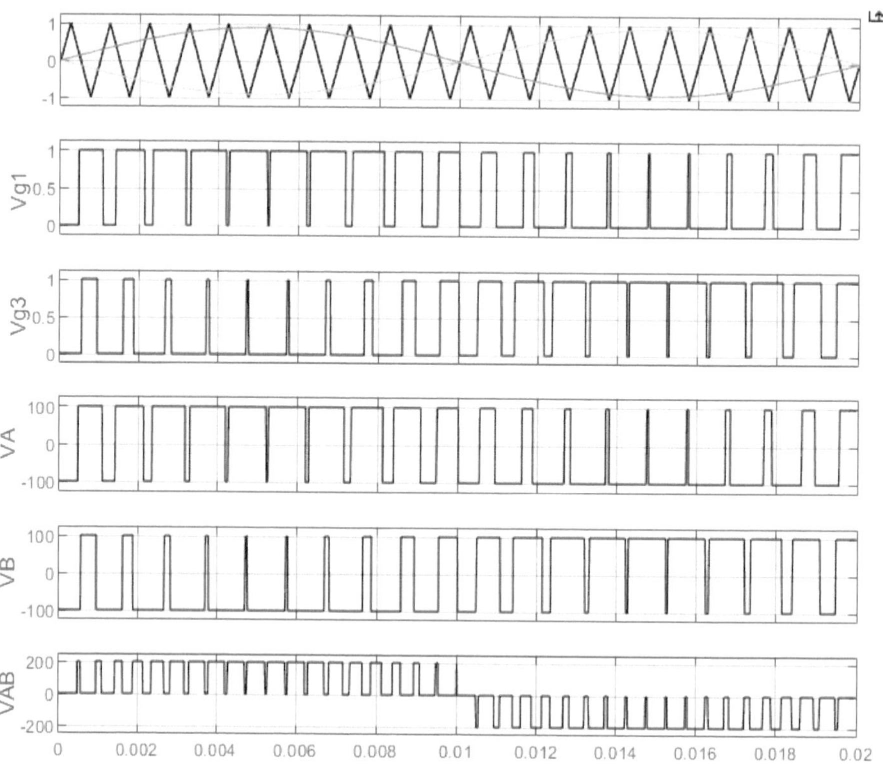

▲ **Fig. 4.29:** Waveforms in the case of unipolar operation

The gating pulses V_{g2} will be the inverse of that of V_{g1} and that of V_{g4} will be the inverse of that of V_{g3}.

4.4.3 Three-Phase PWM Inverter

In a three-phase PWM inverter, there will be three modulating waves, 120° phase shifted with each other, one for every leg. These signals are compared with a common carrier wave and gating pulses for the upper switches in the three legs are generated. The lower switches will be provided with the inverted pulses corresponding to that leg, through dead time generation circuits.

An Arduino programme to generate these three pulses can be written as

```
#include <PWM.h>    // include PWM library:
int32_t frequency = 1000;    //Set freqency

// define variables:
float omega=314.0;
float t;
float out1,out2,out3;

void setup() {
  pinMode (3, OUTPUT);                    //set 3 pin as output :
  pinMode (4,OUTPUT);
  pinMode (5,OUTPUT);
 SetPinFrequencySafe(3, frequency);  // set 3rd pin freuency as 1kHz:
 SetPinFrequencySafe(4, frequency);
 SetPinFrequencySafe(5, frequency);
}
void loop()
{
  t=millis();
out1=sin(omega*t*0.001);
out2=sin(omega*t*0.001-2.094);
out3=sin(omega*t*0.001+2.094);
|
analogWrite(3,out1);
analogWrite(4,out2);
analogWrite(5,out2);

}
```

The carrier wave along with the three modulating waves, gating pulses to the three upper switches, pole voltages and line voltages are depicted in **Fig. 4.30**. The gating signals for the lower switches will be the inverted pulses of corresponding upper switches.

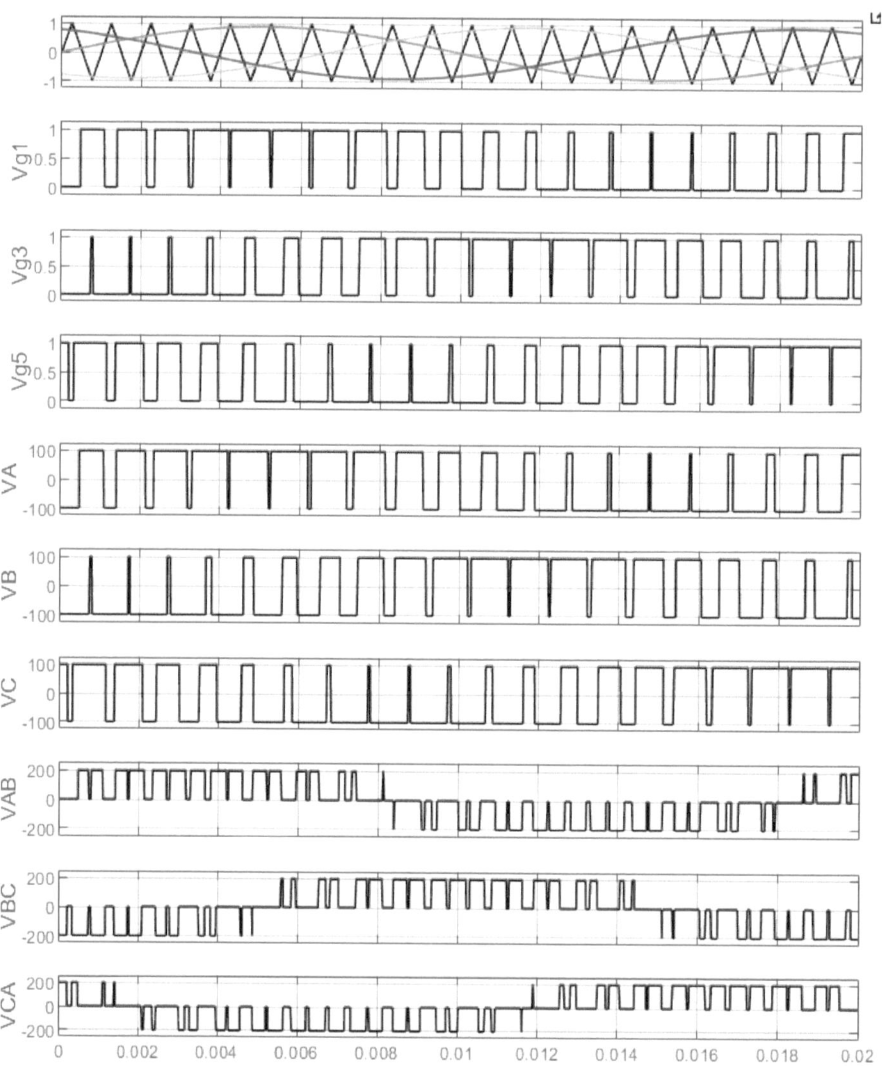

▲ **Fig. 4.30:** Carrier wave, modulating waves, gating pulses, pole voltages and line voltages in a three-phase PWM inverter

05 AC TO DC CONVERTERS

In AC to DC conversion, a fixed AC voltage is converted to a variable DC voltage. For AC inputs, there is a natural zero in voltage occurring twice in every cycle. Hence, there will be a natural zero occurring for the current through the switches. Hence, thyristors are the natural choice for making AC to DC converters. The switching frequency is the supply frequency itself so switching losses will be less. The simplest AC to DC conversion circuit is a half-wave-controlled rectifier.

5.1 Single-Phase Half-Wave-Controlled Rectifier

The circuit diagram of the half-wave rectifier is shown in **Fig. 5.1**. A thyristor switch is connected between a resistive load and the AC source voltage having a peak value of V_m. Since a thyristor is capable of blocking both positive and negative voltages, there will not be any output voltage across the load, until the thyristor is provided with a gating pulse when a positive voltage appears across it. When a gate trigger pulse is applied in the positive half cycle of the supply voltage, the thyristor starts conduction and the supply voltage appears across the load. The thyristor continues its conduction until the current through it becomes zero. With a resistive load, the current will be in phase with the supply voltage and it becomes zero at the zero crossover point of the voltage. Hence, the thyristor becomes automatically OFF when the voltage becomes zero in the positive half-cycle. During the negative half-cycle, the thyristor will not conduct even if it is provided with gating pulses.

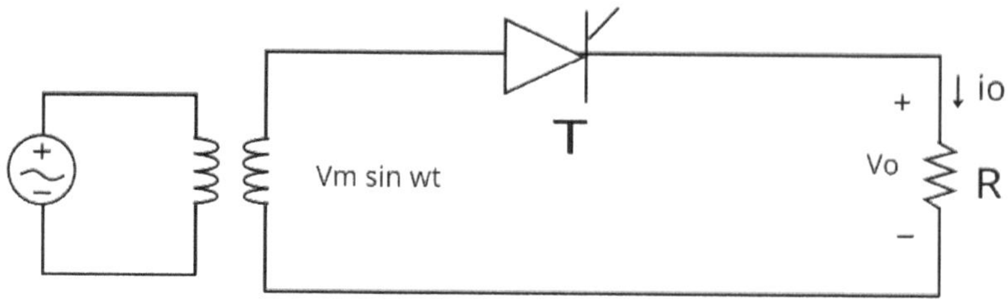

▲ **Fig. 5.1:** Half-wave-controlled rectifier

The input voltage (V_s), triggering pulses (V_g), output voltage (V_o), voltage across the switch (V_T) and current through the switch, which is the same as the load current (I_o), are depicted in **Fig. 5.2**.

The same control circuitry used for triggering IGBTs in the earlier chapter can be used for triggering low-power thyristors. But, since the thyristor is a current-controlled device, the driving circuit must be capable of providing sufficient gate current needed for bringing the thyristor into conduction. If the load is highly inductive, the duration of the triggering pulse should be large enough until the thyristor current becomes larger than the latching current. Generally, the driving pulse will be ANDed with a high-frequency pulse train so that a series of sharp pulses are applied at the gate for a duration sufficient enough to maintain the thyristor conduction. This process is called high-frequency ANDing of the triggering signal. This will help the use of a pulse transformer for isolating control circuitry from the power circuitry. Using Arduino, the high-frequency ANDed signal can be automatically generated by setting a high frequency for the PWM pin used for the generation of triggering pulses. Unlike the case of open-loop operation in DC to DC converters, the triggering pulses are to be synchronised with the supply voltage. The delay between the zero crossing of the supply voltage and the instant at which the triggering pulse is applied measured in terms of the angle of supply voltage is called the delay angle α. By varying this delay angle, the average value of the output voltage can be varied. The DC output voltage can be calculated by finding the average value of the output voltage in terms of α.

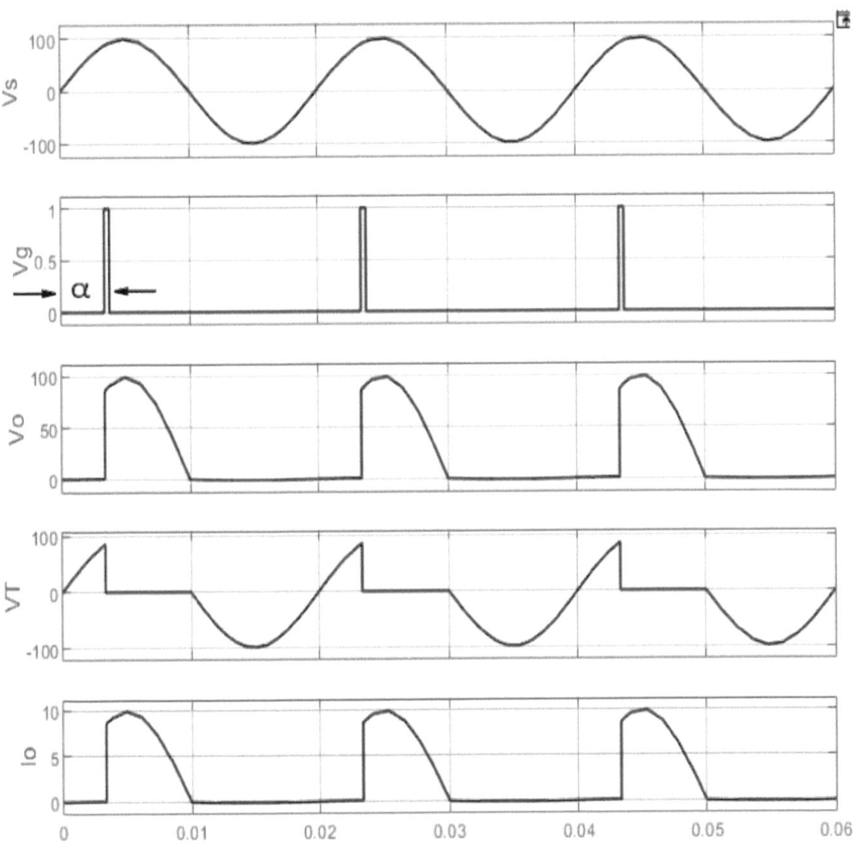

▲ **Fig. 5.2:** Supply voltage, triggering pulses, output voltage, the voltage across the switch and current through the load and thyristor

By varying this delay angle, the average value of the output voltage can be varied. The DC output voltage can be calculated by finding the average value of the output voltage in terms of α.

$$V_{av} = \frac{1}{2\pi}\int_{\alpha}^{\pi} V_m \sin \omega t \, d\omega t \tag{5.1}$$

$$V_{av} = \frac{V_m}{2\pi}(1+\cos \propto) \tag{5.2}$$

The output voltage can be varied from V_m/π to 0 by varying α from 0 to π.

5.1.1 Generation of Gate Pulses Using Arduino.

The triggering pulses are to be generated in synchronism with the supply voltage. The supply voltage is reduced to 3 V using a small measurement transformer and it is converted to a square wave using an op-amp comparator, as shown in **Fig. 5.3**.

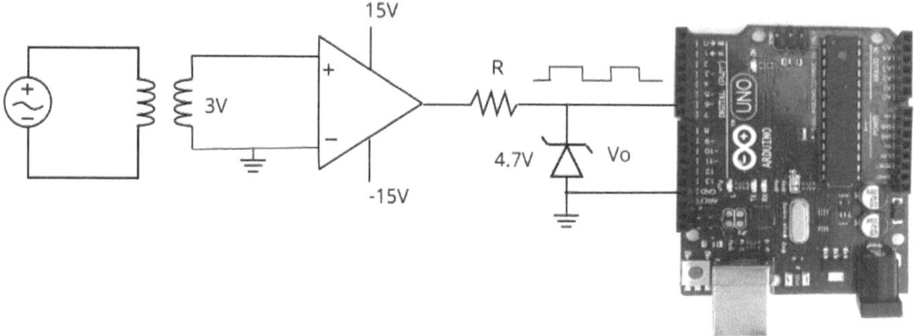

▲ **Fig. 5.3:** Circuit for square wave generation in synchronism with supply voltage

The square wave so generated is given as input to any digital input pin of Arduino. The rising edge of the square wave marks the zero crossing point of the sine wave from which α is to be measured. Hence, the crux of the programme in the generation of triggering pulses is the identification of this point. An easy way is to compare two adjacent sampling points of the square wave and check whether it is negative or not. One gets a negative value at the required zero crossing point only. Once this point is identified, a digital output pin is made at logic 1 for a duration equal to the duration of the triggering pulse, which is equivalent to a high-frequency ANDed signal. Another way to detect the zero crossing point is to continuously check whether the inputted digital signal has become 1 or not. A simple Arduino programme to achieve this is given below:

Programme 5.1

```
int val1;

void setup() {
  pinMode (3, OUTPUT);     //set 3 pin as output :
  pinMode(4,INPUT);
 }
void loop()
{

val1=digitalRead(4);

if ((val1)==1)
{
  delay(2);  // value of firing angle equivalent to 2 millisec
  digitalWrite(3,HIGH);
  delay(1); //pulse width equivalent to 1 millisec
  digitalWrite(3,LOW);
  delay(7); //remaining period in positive cycle
 }
 }
```

The triggering pulses generated in synchronism with the source voltage using the Arduino programme are shown in **Fig. 5.4**, for a firing angle equivalent to 2 milliseconds. The value of the firing angle is 36°.

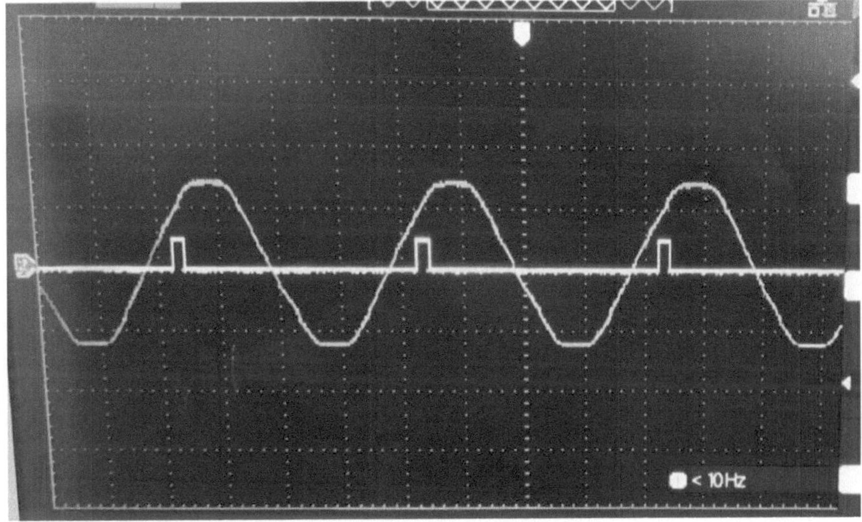

▲ **Fig. 5.4:** Triggering pulses in synchronism with the supply voltage

The output voltages obtained from the converter for a firing angle below 90° and above 90° are shown in **Fig. 7.5**.

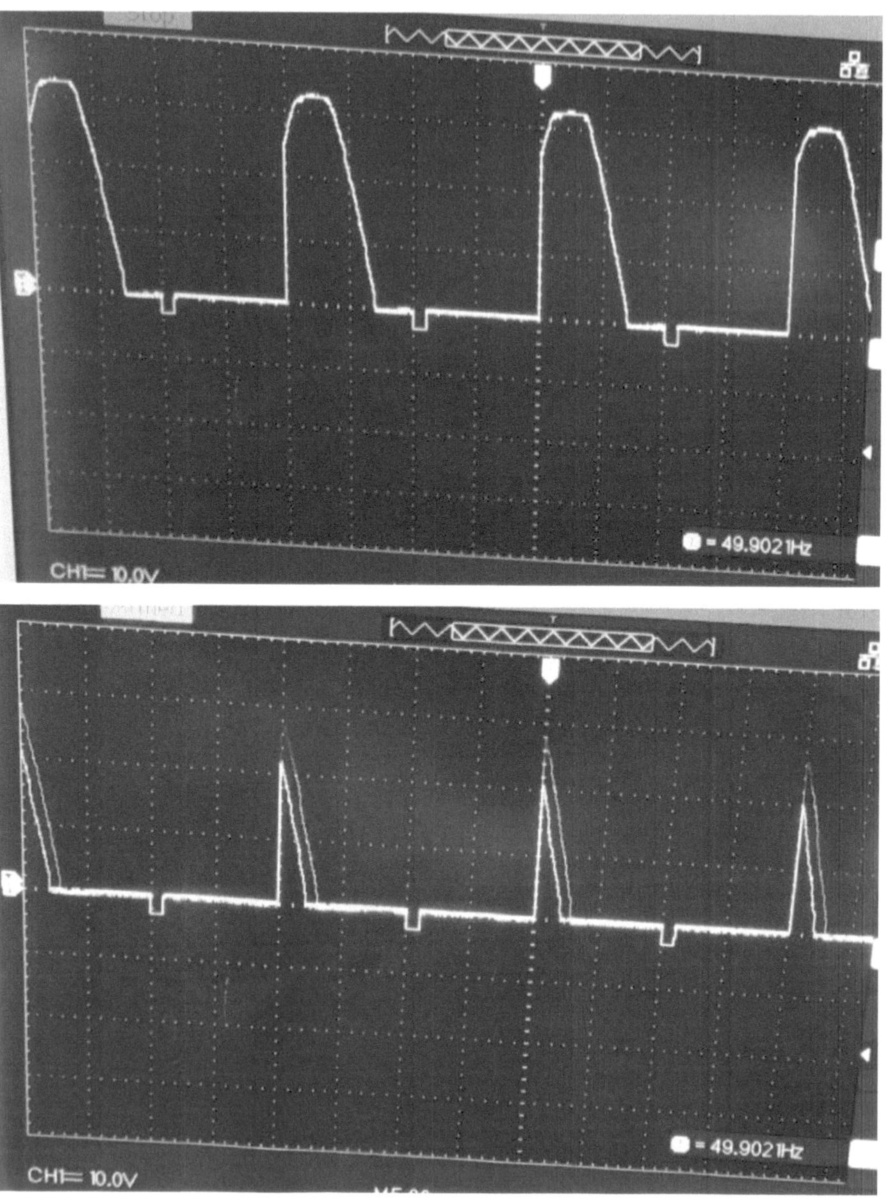

▲ **Fig. 7.5:** Output voltages for firing angles less than and greater than 90°

5.1.2 Half-Wave-Controlled Rectifier with RL Load

When the load is inductive, the current will be lagging the voltage and the current will not be becoming equal to zero along with the voltage becoming zero. Hence, the thyristor will remain conducting beyond the zero crossover point of the voltage until the current becomes zero. Hence, the output voltage will have a negative portion of the supply voltage also, as shown in **Fig. 5.6**. The equation for the average output voltage gets modified accordingly as

$$V_{av} = \frac{1}{\pi} \int_{\alpha}^{\pi+\alpha} V_m \sin \omega t \, d\omega t \tag{5.3}$$

$$V_{av} = \frac{V_m}{\pi} \cos \propto \tag{5.4}$$

The output DC voltage will be equal to V_m/π when $\alpha=0$ as in the case with a resistive load, but it will be equal to 0 at $\alpha=\pi/2$ when the positive side of the output voltage becomes equal to the negative side of the output voltage. Beyond $\alpha=\pi/2$ point, there will be more portion of output voltage on the negative side than on the positive side and hence, the average voltage will be negative for the same direction of current. In this mode, the power, which is the product of voltage and current, is negative and the load will be delivering power to the source. This is possible in the case of an active load like a DC motor. The same Arduino programme used in the previous section can be used in this case as well.

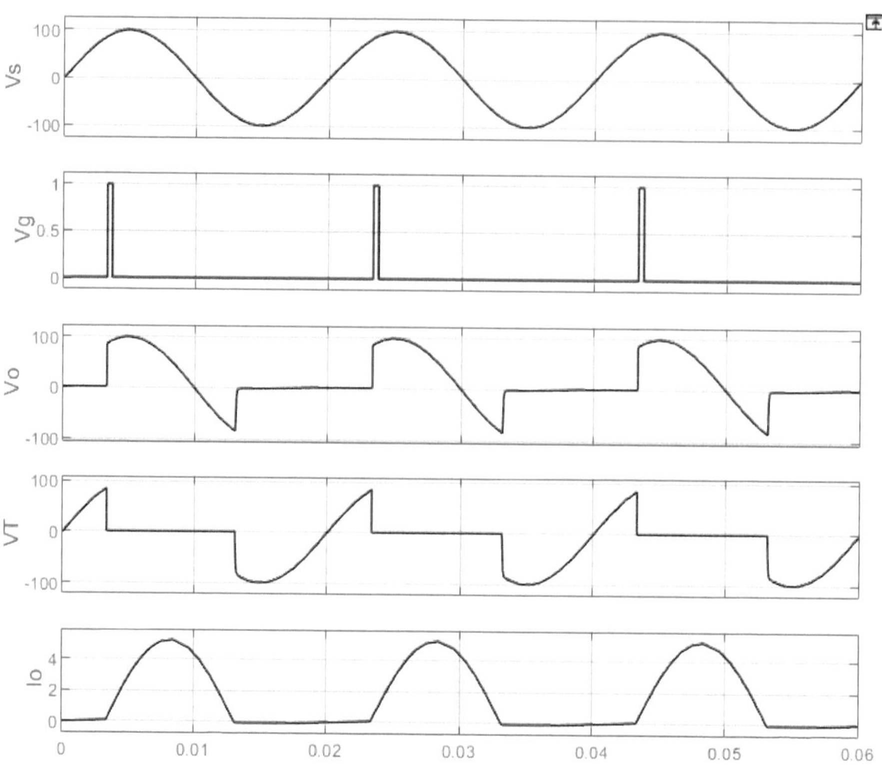

▲ **Fig. 5.6:** Supply voltage, triggering pulses, output voltage, the voltage across the switch and current through the load and thyristor when the load is inductive

5.2 Single-Phase Full-Wave Rectifier

A full-wave-controlled rectifier can be formed in the bridge configuration, as shown in **Fig. 5.7**.

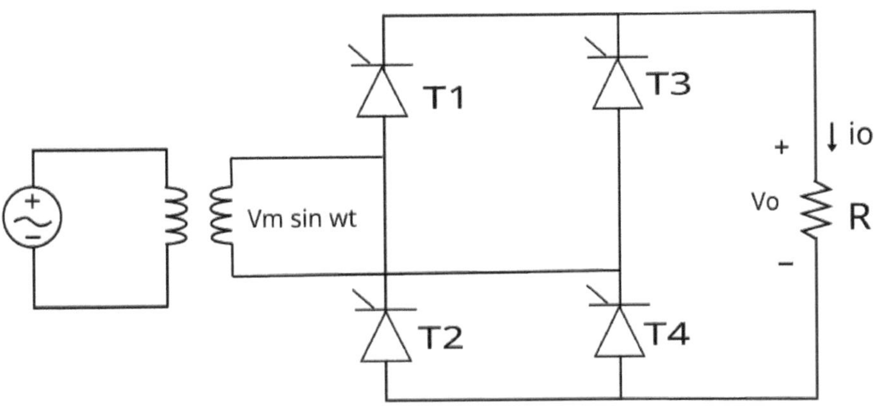

▲ **Fig. 5.7:** Single-phase full-wave rectifier

During the positive half-cycle of the supply voltage, thyristors T1 and T4 are provided with triggering pulses with a delay angle needed. During the negative half-cycle, T2 and T3 are provided with triggering pulses with the same delay. The output voltage waveform along with the supply voltage, triggering pulses and the load current for a resistive load is shown in **Fig. 5.8**.

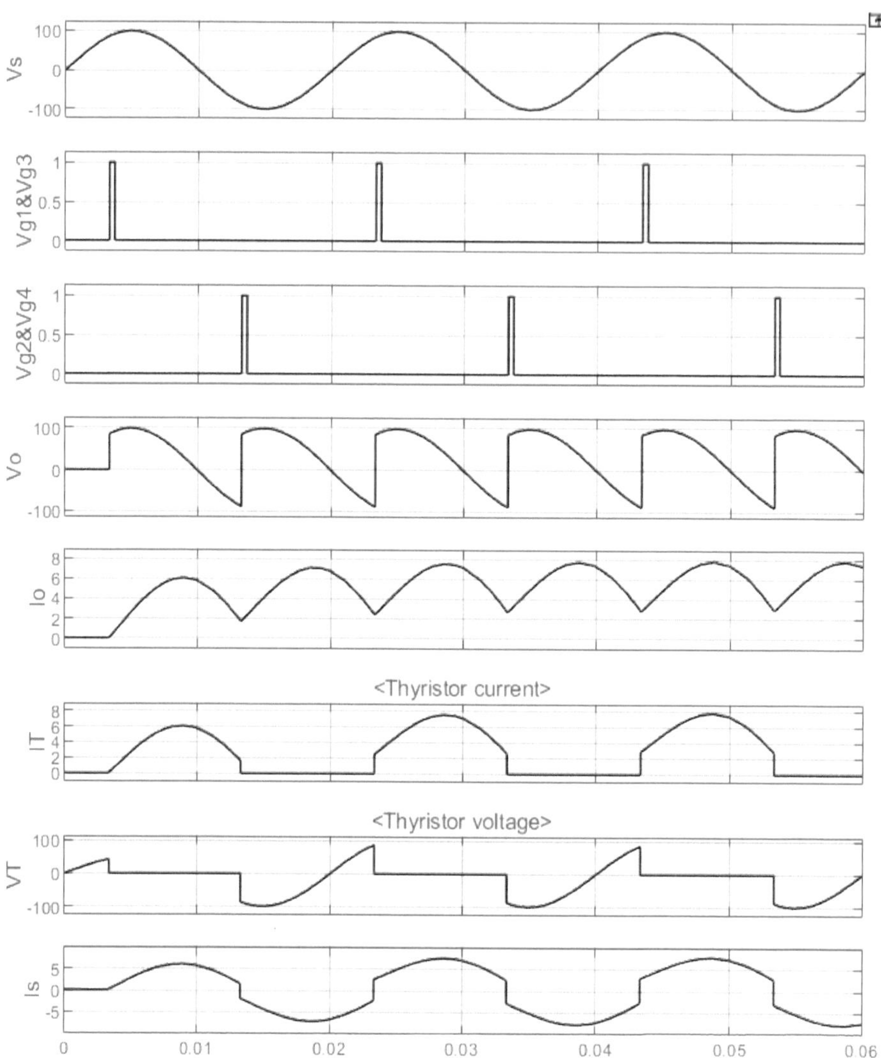

▲ **Fig. 5.8:** Supply voltage, triggering pulses, output voltage, load current, current through thyristor, voltage across thyristor and input current in a full-bridge-controlled rectifier in continuous conduction mode

When the load is inductive, as shown in **Fig. 5.9**, every conducting thyristor will continue conduction beyond the zero crossing point of the supply voltage until the current becomes equal to zero. This instant at which the current becomes equal to zero is called the extinction angle β. This point is decided by the power factor angle of the load. If the other pair of thyristors are triggered before the current in the conducting thyristors reaches the zero point, the operation is said to be in continuous conduction mode. The conducting thyristors will stop conduction before the current zero point and transfer the current to another pair of thyristors. The first case in which the current touches the zero point is called the discontinuous conduction mode. For a given value of inductance, discontinuous conduction will happen if α is more than the power factor angle of the load ϕ, which is given by

$$\varnothing = tan^{-1}\frac{\omega L}{R} \tag{5.5}$$

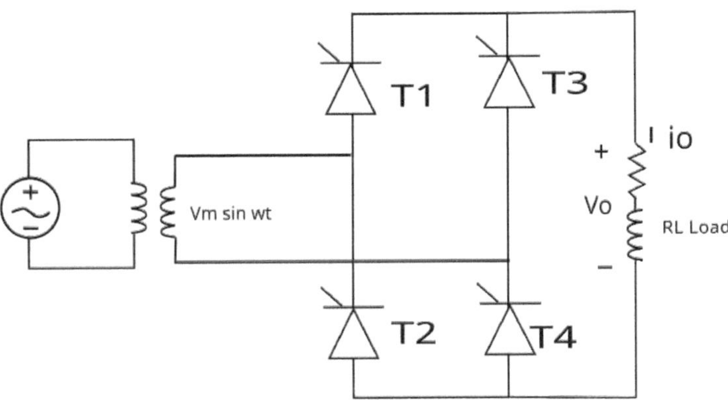

▲ **Fig. 5.9:** Full-wave-controlled rectifier with RL load

The voltage and current waveforms in the continuous conduction mode are shown in **Fig. 5.10**.

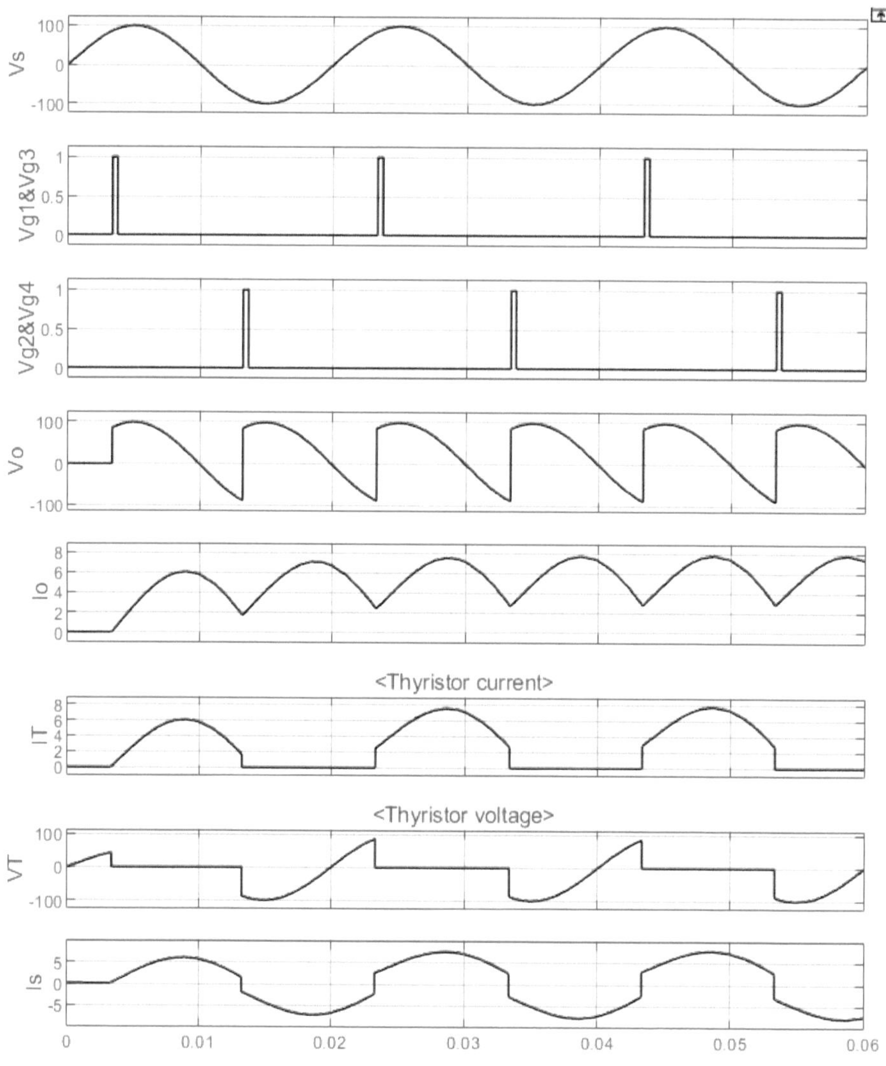

▲ **Fig. 5.10:** Supply voltage, triggering pulses, output voltage, load current, current through thyristor, voltage across thyristor and input current in a full-bridge-controlled rectifier in continuous conduction mode

The voltage and current waveforms when the firing angle is greater than the extinction angle in the discontinuous mode of operation are shown in **Fig. 5.11**.

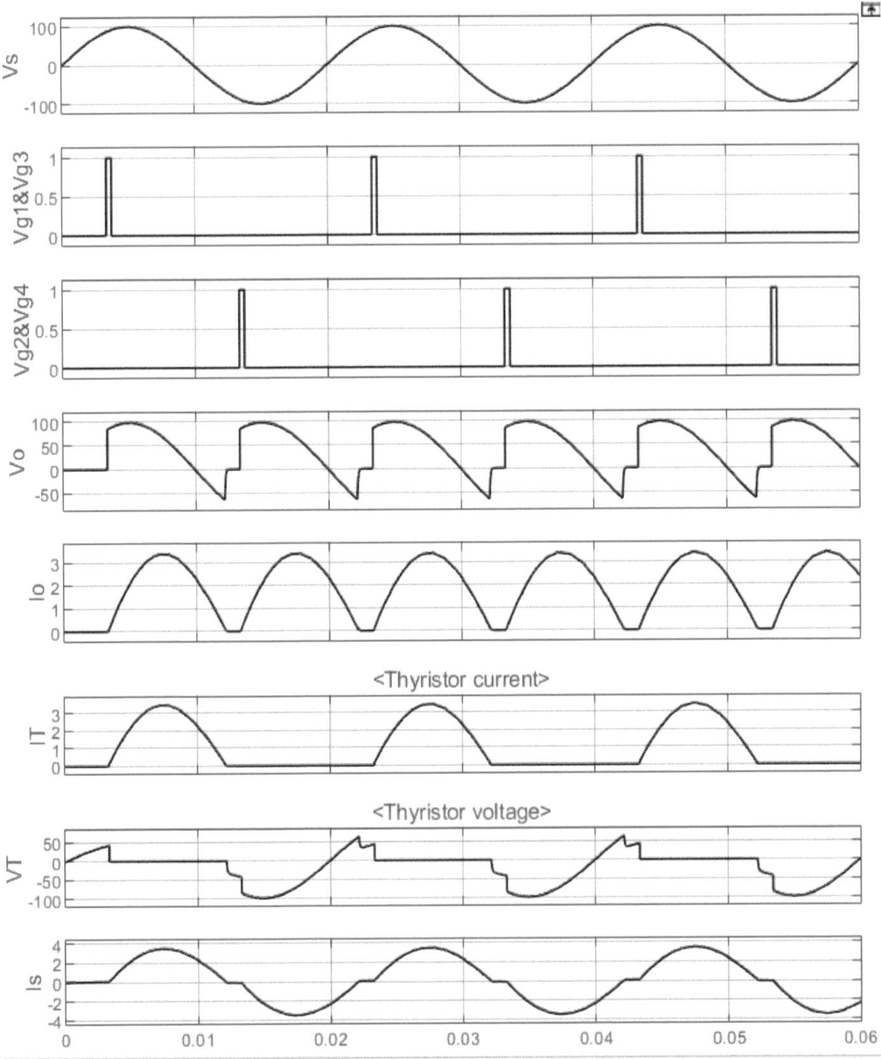

▲ **Fig. 5.11:** Supply voltage, triggering pulses, output voltage, load current, current through thyristor, voltage across thyristor and input current in a full-bridge-controlled rectifier in discontinuous conduction mode

The expression for the average output voltage from a single-phase full-wave-controlled rectifier can be obtained by integrating the input voltage for a period from α to π+α and dividing by π, which is the repetition period of the output waveform.

$$V_{dc} = \frac{1}{\pi} \int_{\alpha}^{\pi+\alpha} V_m \sin \omega t d\omega t \tag{5.6}$$

$$V_{dc} = \frac{2V_m}{\pi} \cos \alpha \tag{5.7}$$

An Arduino programme for the generation of triggering pulses for a full-wave-controlled rectifier is given below:

Programme 5.2

```
int val1;

void setup() {
  pinMode (3, OUTPUT);      //set 3 pin as output :
  pinMode(4,INPUT); // for reading synchronizing square pulses
  pinMode(5,OUTPUT);

 }
void loop()
{

val1=digitalRead(4);

if ((val1)==1) //edge detection
{
  delay(2);  // value of firing angle equivalent to 2 millisec
  digitalWrite(3,HIGH);
  delay(1); //pulse width equivalent to 1 millisec
  digitalWrite(3,LOW);
  delay(7); //remaining period in positive cycle

  delay(2);
  digitalWrite(5,HIGH);
  delay(1);
  digitalWrite(5,LOW);
}
}
```

The pulses generated for triggering thyristors T1 and T4 in the positive half-cycle and the pulses for triggering T2 and T3 in the negative half-cycle are shown in **Fig. 5.12**. Both triggering pulses are shown together in **Fig. 5.13**

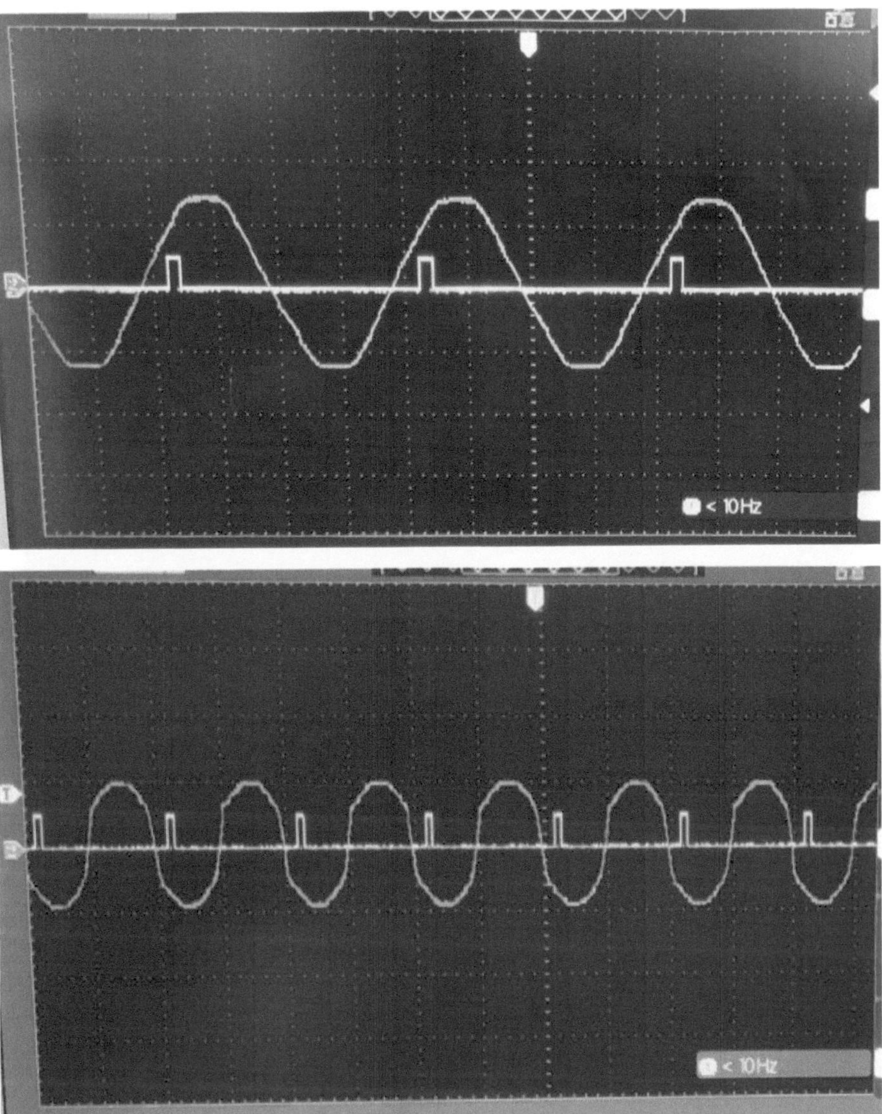

▲ **Fig. 5.12:** Triggering pulses in the positive and negative half-cycles

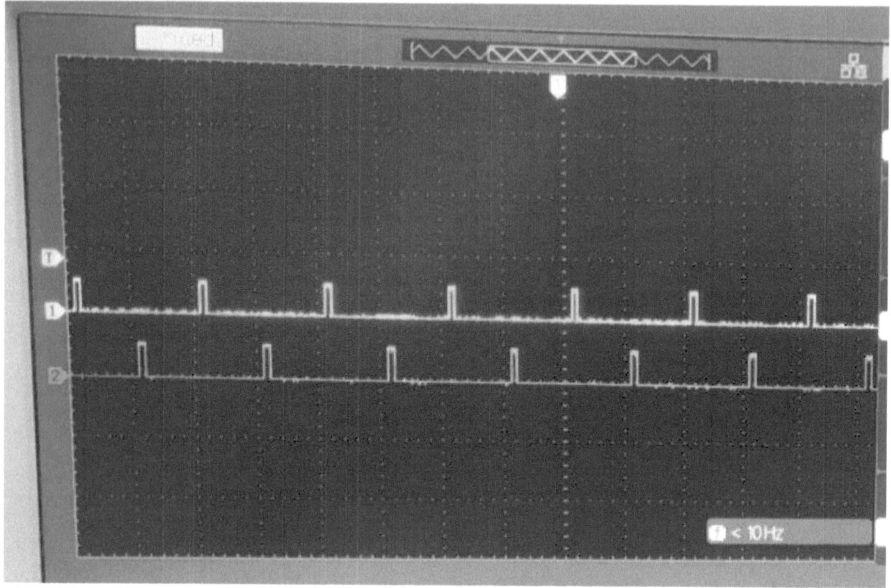

▲ **Fig. 5.13:** Triggering pulses in the positive and negative half-cycles together

5.3 Three-Phase Half-Wave-Controlled Converter

The working of a three-phase controlled converter can be easily understood by considering a three-phase uncontrolled half-wave converter. In an uncontrolled converter, thyristors will be replaced with diodes. Diodes can be considered as thyristors operating at $\alpha = 0$. The circuit diagram of an uncontrolled three-phase half-wave converter is shown in **Fig. 5.14**.

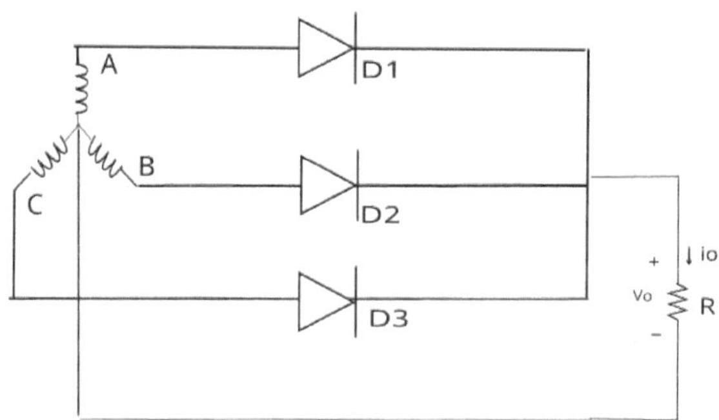

▲ **Fig. 5.14:** Three-phase half-wave uncontrolled rectifier

When any one of the diodes is conducting, the voltage across other diodes will be line-to-line. Hence, diodes will start conduction at the zero crossing point of the line-to-line voltages. D1 will conduct when phase A voltage is more positive than the other voltages. D2 will conduct when phase B voltage is more positive than the other two voltages and D3 will conduct when phase C voltage is more positive, as shown in **Fig. 5.15**.

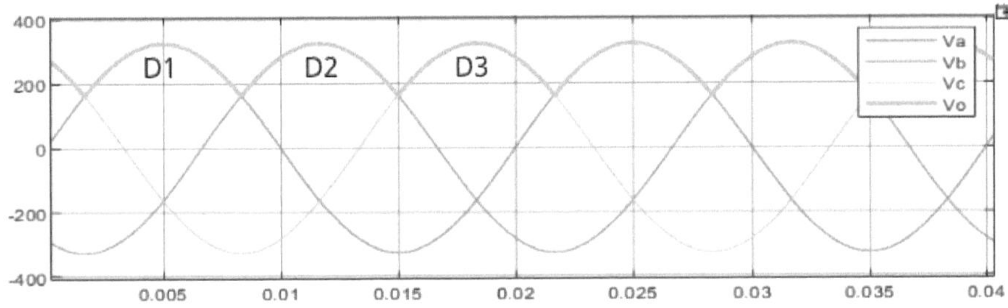

▲ **Fig. 5.15:** Phase voltages and output voltage from a half-bridge-uncontrolled converter

It can be observed that if we replace the diodes with thyristors, as shown in **Fig. 5.16**, α=0 point for T1 will be 30° lagging from the zero crossover point of phase A voltage. T1 will conduct for 120° and T2 will start conducting thereafter. Hence, the α=0 point for T2 will be 30°+120° = 150° from the zero crossing point of phase A. Similarly, the α=0 point for T3 will be 30°+240°=270° from the zero crossing point of phase A. Hence, by monitoring only one phase voltage, the triggering pulses for all phases can be generated. A programme for generating triggering pulses for a three-phase-controlled rectifier is shown below. The supply voltage waveforms, triggering pulses, output voltage, output current, current through a thyristor and voltage across it are shown in **Fig. 5.17**, for α=0.

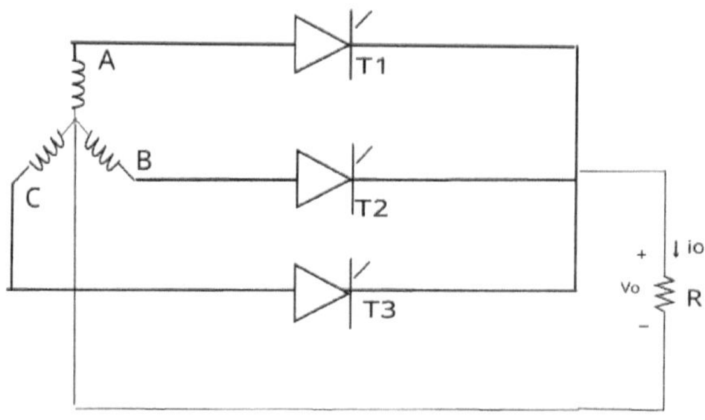

▲ **Fig. 5.16:** Three-phase half-wave-controlled rectifier

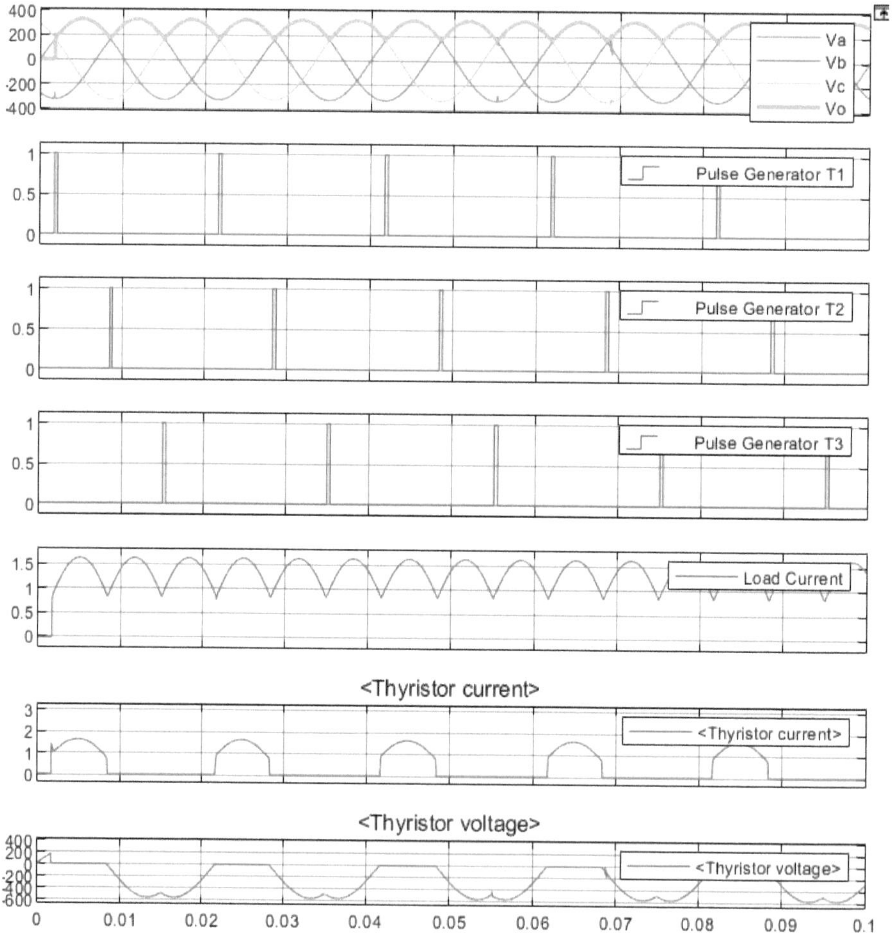

▲ **Fig. 5.17:** Voltage and current waveforms of a three-phase half-wave-controlled rectifier for α=0

The waveforms for α=60° is shown in **Fig. 5.18**

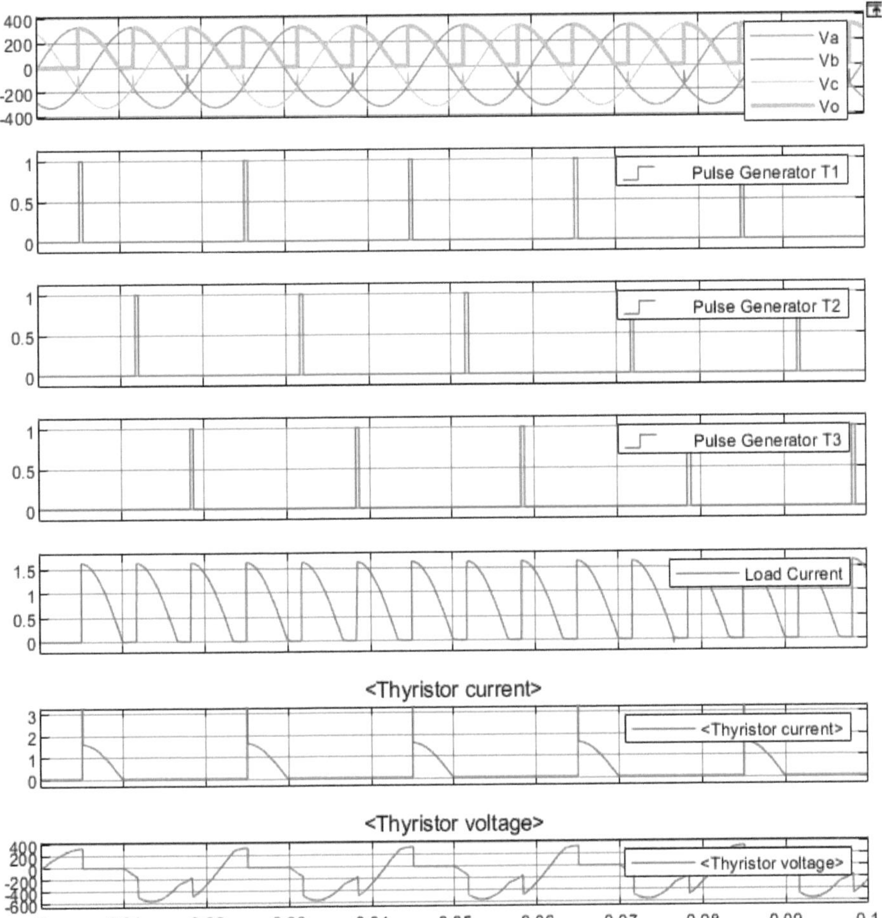

▲ **Fig. 5.18:** Voltage and current waveforms of a half-wave-controlled rectifier for α=0

The same waveforms for a highly inductive load working in continuous conduction mode are shown in **Fig. 5.19**.

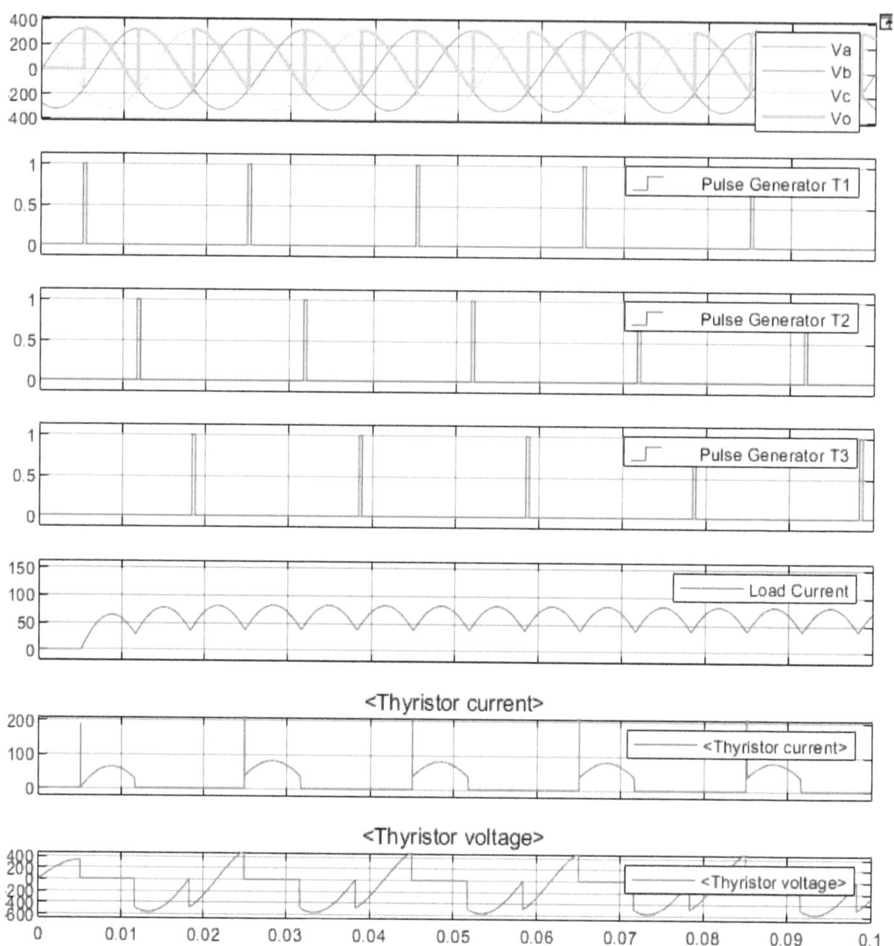

▲ **Fig. 5.19:** Waveforms with highly inductive load in continuous conduction mode of operation

For the implementation of the three-phase half-wave-controlled rectifier, one has to generate three triggering pulses at angles 30°+α, 150°+ α and 270°+ α. The Arduino programme written for a single-phase full-wave-controlled rectifier can be slightly modified for this purpose. Instead of generating two pulses for a single-phase rectifier, one has to generate three pulses. For more accuracy, delays are given in microseconds rather than in milliseconds, here. The programme is given below.

Programme 5.3

```
int val1;
int initial_delay;
int alpha=0; // value for delay angle in microseconds

void setup() {
  pinMode (3, INPUT); //for reading square pulses
  pinMode(4,OUTPUT); // for triggering T1
  pinMode(5,OUTPUT); // for triggering T2
  pinMode(6,OUTPUT); // for triggering T3
  initial_delay = 1700+alpha;//delay for 30+alpha

 }
void loop()
{

val1=digitalRead(3);

if ((val1)==1) //edge detection
{
  delayMicroseconds(initial_delay);  // delay for 30+alpha
  digitalWrite(4,HIGH);
  delay(1); //pulse width equivalent to 1 millisec
  digitalWrite(4,LOW);
  delayMicroseconds(5700); //remaining period in 120 degrees


  digitalWrite(5,HIGH);
  delay(1);
  digitalWrite(5,LOW);
  delayMicroseconds(5700);
  |
  digitalWrite(6,HIGH);
  delay(1);
  digitalWrite(6,LOW);
}
}
```

The expression for the average output voltage from a three-phase half-wave-controlled rectifier can be obtained by integrating the input voltage for a period from $30^{\circ}+\alpha$ to $150^{\circ}+\alpha$ and dividing by π, which is the repetition period of the output waveform.

$$V_{dc} = \frac{3}{2\pi} \int_{\frac{\pi}{6}+\alpha}^{\frac{5\pi}{6}+\alpha} V_m \sin \omega t d\omega t \tag{5.8}$$

$$V_{dc} = \frac{3\sqrt{3}V_m}{2\pi} \cos \alpha \tag{5.9}$$

5.4 Three-Phase Full-Wave-Controlled Rectifier

A three-phase full-wave-controlled rectifier can be considered as two half-wave rectifiers connected back-to-back, as shown in **Fig. 5.20**. Switch T1 will be forward-biased when phase voltage V_a is more positive with respect to other voltages. Similarly, T2 and T3 will be forward-biased when V_b and V_c, respectively, are more positive with respect to other voltages. In a similar manner, T4 will be forward-biased when V_a is more negative than the other voltages. T5 and T2 will be conducting when V_b and V_c, respectively are more negative than other voltages. All the thyristors can be triggered when they are forward-biased, with an appropriate delay.

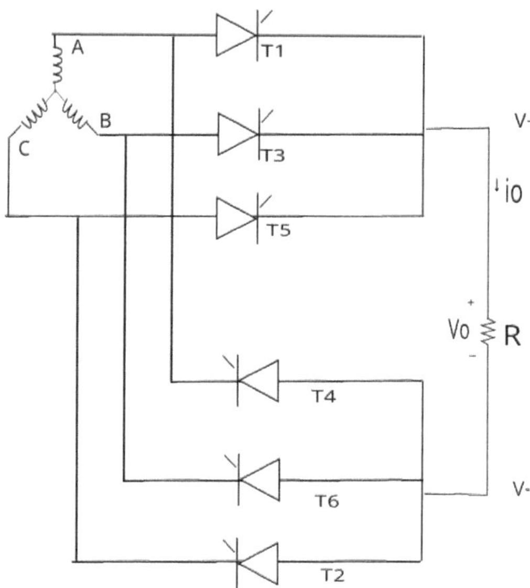

▲ **Fig. 5.20:** Three-phase full-wave rectifier considered as two half-wave rectifiers

Voltages at the upper and lower terminals of the load (V+ and V−) are shown in **Fig. 5.21**. The load voltage is the difference between V+ and V− and it is also shown in **Fig. 5.21**. It can be observed that the load voltage is formed as parts of line-to-line voltages, changing phases in every 60°. The α=0 point of T1 will be 30° away from the zero crossing point of V_a. The α=0 point for T2 will be 60° away from that of T1. Similarly, T3 has to be triggered after 60° from the point of triggering of T2 and so on.

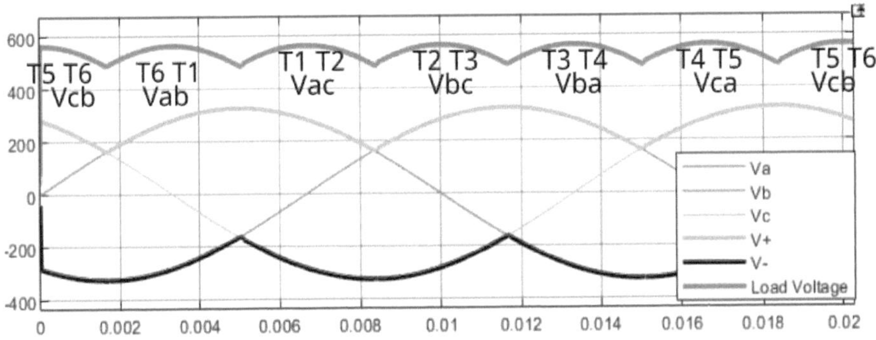

▲ **Fig. 5.21:** Load voltage along with phase voltages

Normally, three-phase rectifiers will be drawn like a bridge circuit, as shown in **Fig. 5.22**

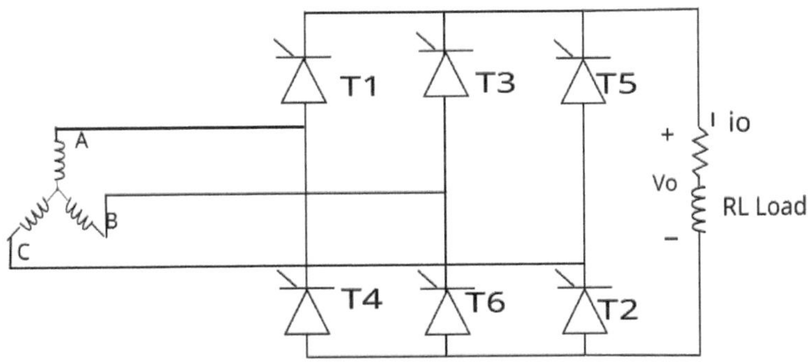

▲ **Fig. 5.22:** Three-phase full-wave rectifier drawn in standard form

The triggering pulses for a three-phase-controlled rectifier can be generated using Arduino by modifying the programme written for the half-wave converter. A pulse has to be generated in every 60° from the 'α=0' point. 'α=0' point can be located 30° away from the zero crossing point of phase A voltage. Hence, only one voltage needs to be measured for the generation of firing pulses for all the six switches. A programme for achieving this is shown below:

Programme 5.4

```
int initial_delay;
int alpha=1700; // value for delay angle

void setup() {
  pinMode (3, INPUT); // reading square pulses
  pinMode(4,OUTPUT); // for triggering T1
  pinMode(5,OUTPUT); // for triggering T2
  pinMode(6,OUTPUT); // for triggering T3
  pinMode(7,OUTPUT); // for triggering T3
  pinMode(8,OUTPUT); // for triggering T3
  pinMode(9,OUTPUT); // for triggering T3
  initial_delay=1700+alpha;
   }
void loop()
{
val1=digitalRead(3);
if ((val1)==1) //edge detection
{
  delayMicroseconds(initial_delay);
  digitalWrite(4,HIGH);
  delay(1); //pulse width equivalent to 1 millisec
  digitalWrite(4,LOW);
  delayMicroseconds(2300); //remaining period in 60 degrees

 digitalWrite(5,HIGH);
 delay(1);
 digitalWrite(5,LOW);
 delayMicroseconds(2300);

 digitalWrite(6,HIGH);
 delay(1);
 digitalWrite(6,LOW);
 delayMicroseconds(2300);
```

```
    digitalWrite(7,HIGH);
    delay(1);
    digitalWrite(7,LOW);
    delayMicroseconds(2300);

    digitalWrite(8,HIGH);
    delay(1);
    digitalWrite(8,LOW);
    delayMicroseconds(2300);

    digitalWrite(9,HIGH);
    delay(1);
    digitalWrite(9,LOW);
}
}
```

The load voltage, load current, source current, the voltage across a thyristor and current through a thyristor for α=0 and α=60° for an RL load are shown in **Fig. 5.23** and 5.24, respectively.

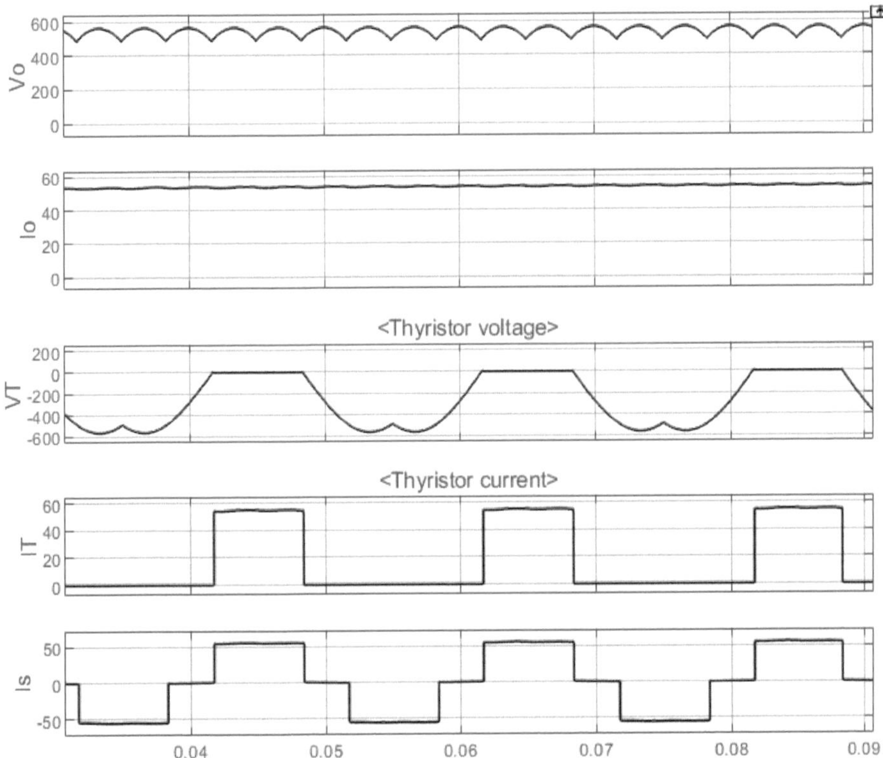

▲ **Fig. 5.23:** Waveforms of three-phase full-wave rectifier for α=0

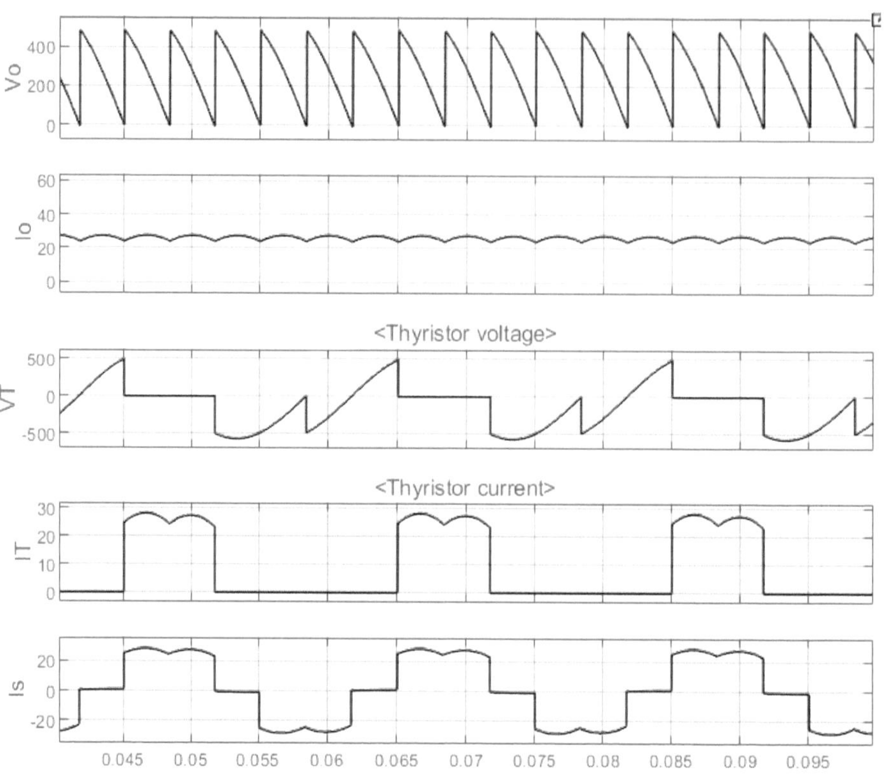

▲ **Fig. 5.24:** Waveforms of three-phase full-wave rectifier for α=60°

Since the output voltage gets repeated every 60°, the average output voltage can be evaluated by integrating any line-to-line voltage for a duration of 60°.

$$V_{dc} = \frac{3}{\pi} \int_{\frac{\pi}{6}+\alpha}^{\frac{\pi}{2}+\alpha} V_{ab} d(\omega t) \tag{5.10}$$

$$V_{dc} = \frac{3}{\pi} \int_{\frac{\pi}{6}+\alpha}^{\frac{\pi}{2}+\alpha} \sqrt{3} V_m \sin(\omega t + \frac{\pi}{6} d(\omega t) \tag{5.11}$$

$$V_{dc} = \frac{3\sqrt{3}}{\pi} V_m \, cos\alpha \tag{5.12}$$

5.4 Half-Controlled Rectifier

In a half-controlled rectifier, the lower thyristor in every leg is replaced with a diode. In the case of a fully controlled converter, the conducting thyristors will continue their conduction until the inductive load current becomes zero or until the next pair of thyristors is triggered. Since a diode cannot conduct when the voltage across it becomes negative, there will not be a current path through a thyristor in the negative half-cycle even when the load is inductive. Hence, thyristors stop conduction at the voltage zero crossing point itself. But, the current through an inductor cannot be made zero instantaneously; a current path has to be provided for the load current to get decayed to zero. This is done by connecting a freewheeling diode in parallel with the load. The circuit diagram and waveforms of a single-phase half-controlled rectifier are shown in **Fig. 5.25** and 5.26, respectively.

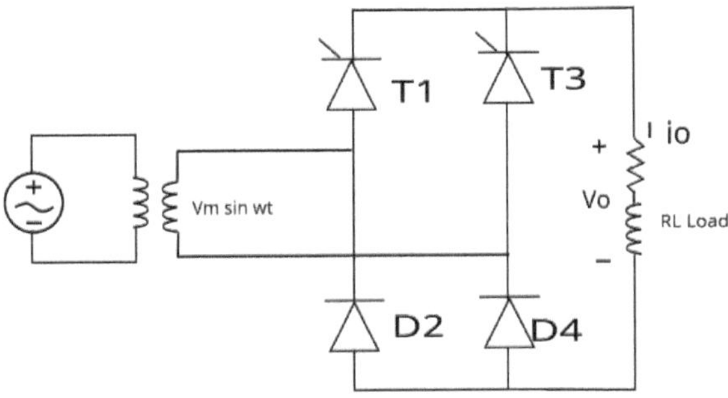

▲ **Fig. 5.25:** Single-phase full-wave half-controlled rectifier

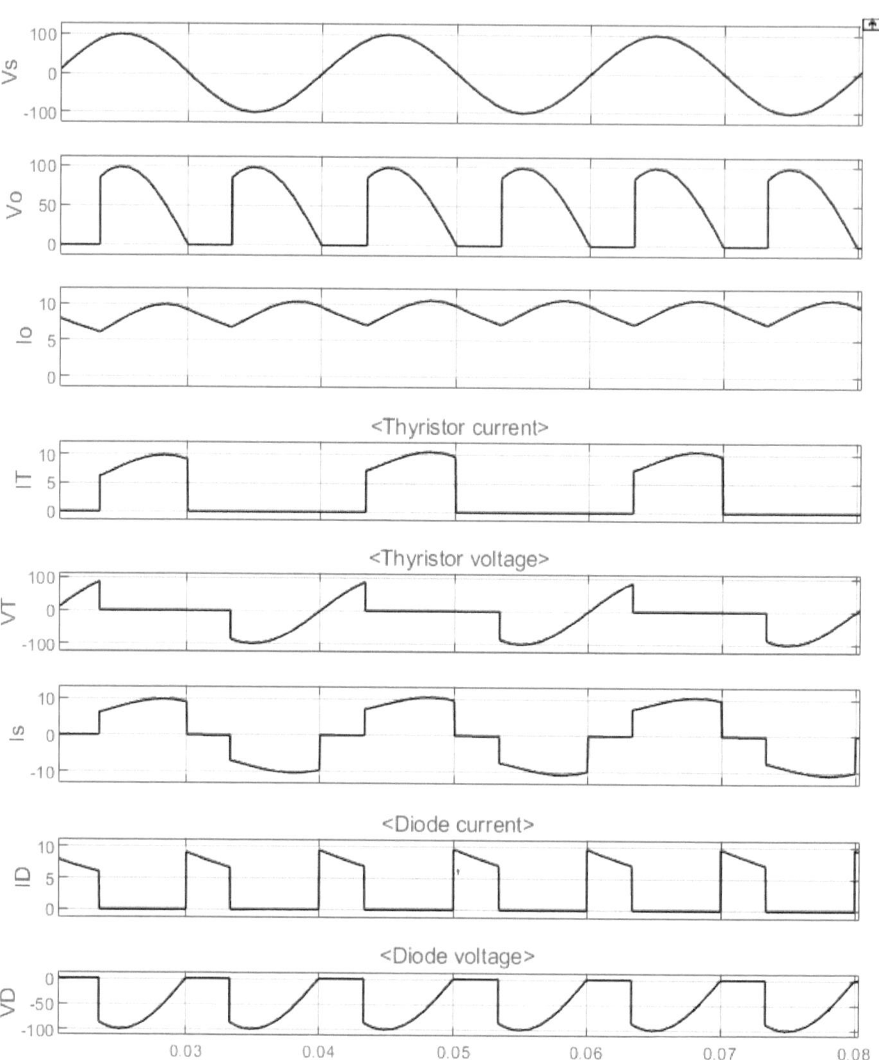

▲ **Fig. 5.26:** Output voltage, output current, source current and device voltages and currents of a single-phase half-controlled converter

The Arduino programme used for triggering a fully controlled full-wave converter can be used for triggering this circuit as well. The programme is reproduced below:

Programme 5.5

```
int val1;

void setup() {
  pinMode (3, OUTPUT);      //set 3 pin as output :
  pinMode(4,INPUT); // for reading synchronizing square pulses
  pinMode(5,OUTPUT);

 }
void loop()
{

val1=digitalRead(4);

if ((val1)==1) //edge detection
{
   delay(2);  // value of firing angle equivalent to 2 millisec
   digitalWrite(3,HIGH);
   delay(1); //pulse width equivalent to 1 millisec
   digitalWrite(3,LOW);
   delay(7); //remaining period in positive cycle

   delay(2);
   digitalWrite(5,HIGH);
   delay(1);
   digitalWrite(5,LOW);
}
}
```

Here, only two pulses for T1 and T3 are needed to be applied. No triggering is needed for lower devices in both the legs. Triggering pulses for T1 will be available at pin3 and that for T3 will be available at pin5.

A three-phase half-controlled rectifier can be formed by adding one more leg to a single-phase half-controlled rectifier. The circuit diagram and the waveforms are shown in **Fig. 5.27** and 5.28, respectively.

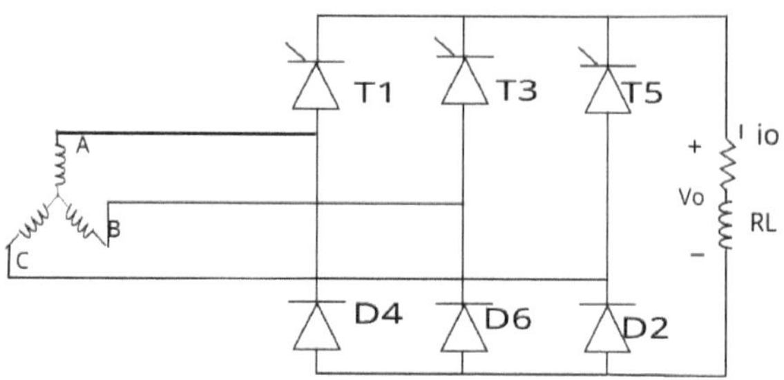

▲ **Fig. 5.27:** Three-phase half-controlled rectifier

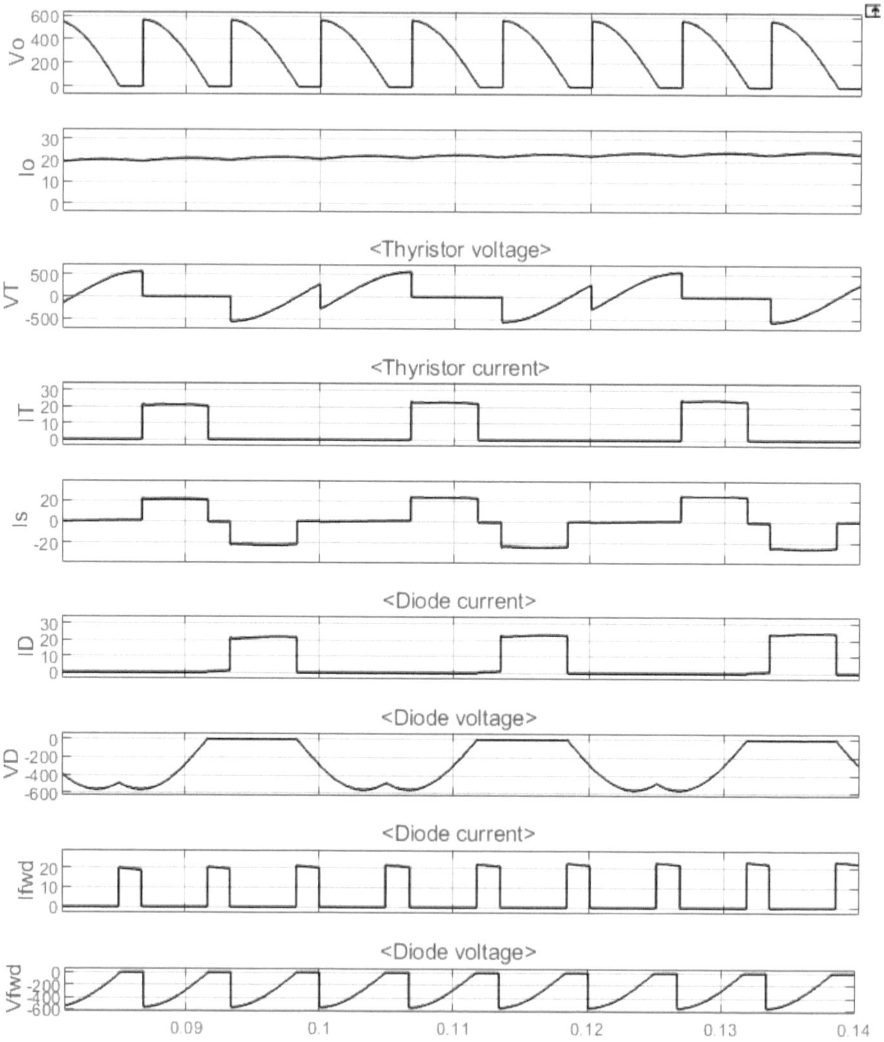

▲ **Fig. 5.28:** Waveforms of a three-phase half-controlled rectifier

The firing pulses for a three-phase half-controlled rectifier can be generated by slightly modifying the Arduino programme written for a three-phase fully controlled rectifier, by removing the pulses for the lower switches. The programme will be the same as that written for a half-wave rectifier, which is reproduced below:

Programme 5.6

```
int val1;
int initial_delay;
int alpha=0; // value for delay angle in microseconds

void setup() {
  pinMode (3, INPUT); //for reading square pulses
  pinMode(4,OUTPUT); // for triggering T1
  pinMode(5,OUTPUT); // for triggering T2
  pinMode(6,OUTPUT); // for triggering T3
  initial_delay = 1700+alpha;//delay for 30+alpha

 }
void loop()
{

val1=digitalRead(3);

if ((val1)==1) //edge detection
{
  delayMicroseconds(initial_delay);   // delay for 30+alpha
  digitalWrite(4,HIGH);
  delay(1); //pulse width equivalent to 1 millisec
  digitalWrite(4,LOW);
  delayMicroseconds(5700); //remaining period in 120 degrees


  digitalWrite(5,HIGH);
  delay(1);
  digitalWrite(5,LOW);
  delayMicroseconds(5700);

  digitalWrite(6,HIGH);
  delay(1);
  digitalWrite(6,LOW);
}
}
```

06 AC TO AC CONVERSION

There are two types of AC to AC conversion. In the first case, an AC voltage of fixed magnitude and frequency is transformed into another AC voltage of variable magnitude of the same frequency. These types of circuits are called voltage controllers. The second category of circuits converts voltages of fixed magnitude and frequency to AC voltages of variable magnitude and variable frequency. Such circuits are called cyclo converters. The number of switches needed in cyclo conversion is notably higher than that needed in the case of voltage controllers. In cyclo conversion, the output voltage would generally be carved out from one or more rectifier circuits, operated with sinusoidally varying delay angles. The wave shape will be far different from a sine wave. The control of such a circuit is very complex as well. Since it is easy to get a variable voltage frequency wave from an inverter, cyclo converters are not used widely. This requirement is met by converting AC into DC using uncontrolled rectification using diodes, and this fixed DC is converted to an AC voltage of the required magnitude and frequency using inverters. Hence, voltage-controller circuits alone will be discussed in this book.

6.1 Single-Phase Voltage Controller

In a single-phase voltage converter, two back-to-back thyristors are connected in between the supply and load, as shown in **Fig. 6.1**. Both thyristors are triggered with a delay angle in their respective forward-biased conditions.

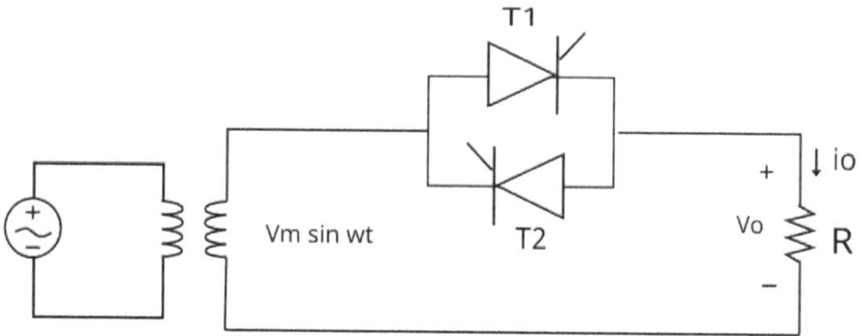

▲ **Fig. 6.1:** Single-phase controller

T1 is triggered in the positive half-cycle and T2 is triggered in the negative half-cycle. For a resistive load, the voltage and current waveforms will be identical so that the current reaches zero at the same moment when the voltage reaches zero. Hence, both thyristors stop conduction at their voltage zero crossing points. This is depicted in **Fig. 6.2**.

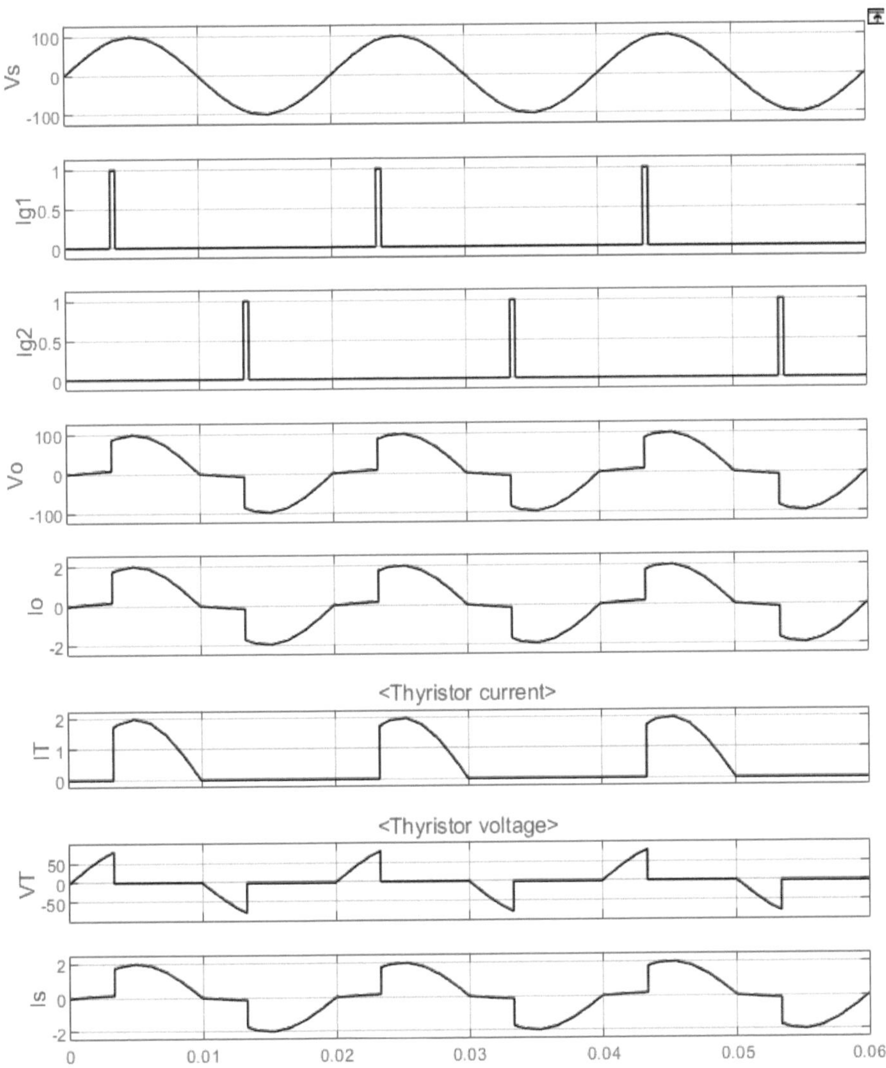

▲ **Fig. 6.2:** Supply voltage, triggering pulses, output voltage, output current, thyristor voltage, thyristor current and input current for a resistive load

When the load is inductive, the current lags the voltage and the thyristor T1 continues to conduct beyond the voltage zero point in the positive half-cycle. T1 will turn off when the current becomes zero. This point is called the extinction angle, β, which is dependent on the power factor angle of the load. This process happens with T2 as well in the negative half-cycle. Since T1 continues to conduct up to β, the voltage across both switches will continue to be zero up to this point. Hence, any attempt to turn on T2 during this period will fail. T1 and T2 can be successfully triggered only beyond the β point in both directions. Hence, control action in a voltage controller will happen only if $\alpha > \beta$.

Waveforms for the case of an inductive load working with a delay angle greater than the power factor angle are shown in **Fig. 6.3**.

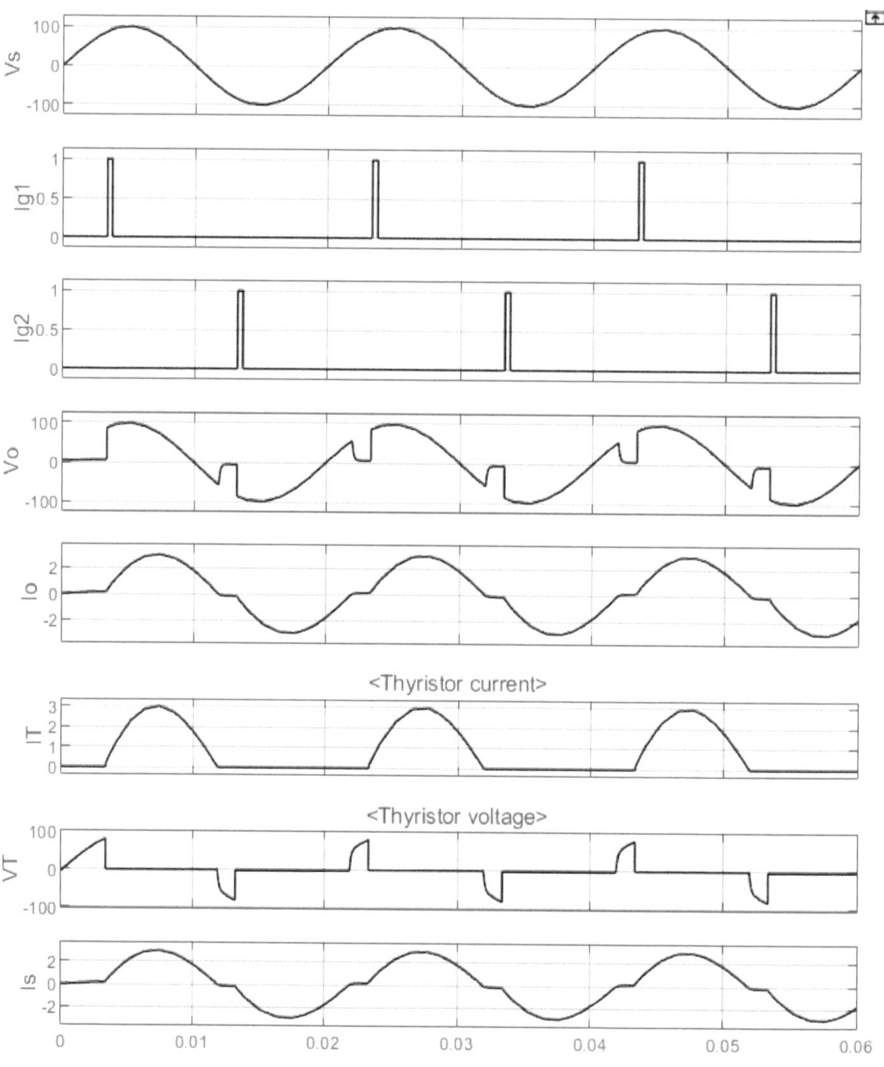

▲ **Fig. 6.3:** Waveforms of a single-phase voltage controller with inductive load

In order to implement a single-phase voltage controller, two triggering pulses are to be generated in synchronism with the supply voltage. The control circuitry and the programme are the same as those with a single-phase full-wave rectifier. The supply voltage is reduced to a low-voltage signal and given to a comparator op-amp circuit. The output will be a square wave in sync with the supply voltage. This signal is given as an input signal at a digital pin in the Arduino board for edge detection. Once the zero crossing edge is detected, pulses are outputted at two digital output ports after proper delays. The delay for T1 will be the time period corresponding to α and that for T2 will be the time period corresponding to 180°+α. A programme in which digital pin 4 is used as an input pin and pins 3 and 5 are used as output pins for generating trigger pulses is given below.

Programme 6.1

```
int val1;

void setup() {
  pinMode (3, OUTPUT);     //set 3 pin as output :
  pinMode(4,INPUT); // for reading synchronizing square pulses
  pinMode(5,OUTPUT);

 }
void loop()
{

val1=digitalRead(4);

if ((val1)==1) //edge detection
{
  delay(2);  // value of firing angle equivalent to 2 millisec
  digitalWrite(3,HIGH);
  delay(1); //pulse width equivalent to 1 millisec
  digitalWrite(3,LOW);
  delay(7); //remaining period in positive cycle

  delay(2);
  digitalWrite(5,HIGH);
  delay(1);
  digitalWrite(5,LOW);
}
}
```

Since the output voltage from voltage controllers is AC, it is evaluated as RMS voltage. The RMS value is evaluated by integrating the square of the waveform over the repetition period. It is then divided by the period of repetition. The square root of this value will be the RMS value. In this case, the RMS value can be evaluated as

$$V_{rms}^2 = \frac{1}{\pi}\int_{\alpha}^{\pi}(V_m \sin\omega t)^2\, d\omega t \tag{6.1}$$

$$V_{rms} = \frac{V_m}{\pi}\sqrt{\left[\frac{1}{\pi}\left(\pi - \alpha + \frac{\sin 2\alpha}{2}\right)\right]} \tag{6.2}$$

The output voltage can be varied from 0 to the maximum of supply voltage by varying α from 0 to 180°.

6.2 Three-Phase Voltage Controller

A three-phase voltage controller can be formed by inserting back-to-back connected thyristors in all the three phases. The triggering pulses are applied at the appropriate positive and negative half-cycles of each phase. The numbering of thyristors is done in the sequence in which each thyristor is fired in a cycle. It can be observed that the sequence of triggering is the same as with a three-phase full-wave-controlled rectifier. The diagram of a full-wave controller is given in **Fig. 6.4**.

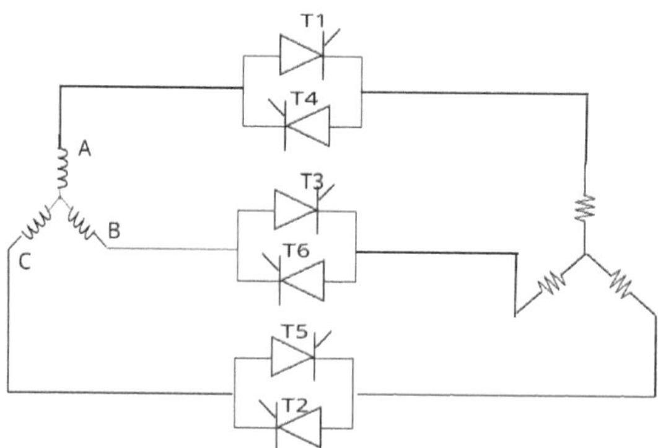

▲ **Fig. 6.4:** Three-phase voltage controller

The voltage and current waveforms associated with this controller pertaining to phase A for an inductive load are depicted in **Fig. 6.5**.

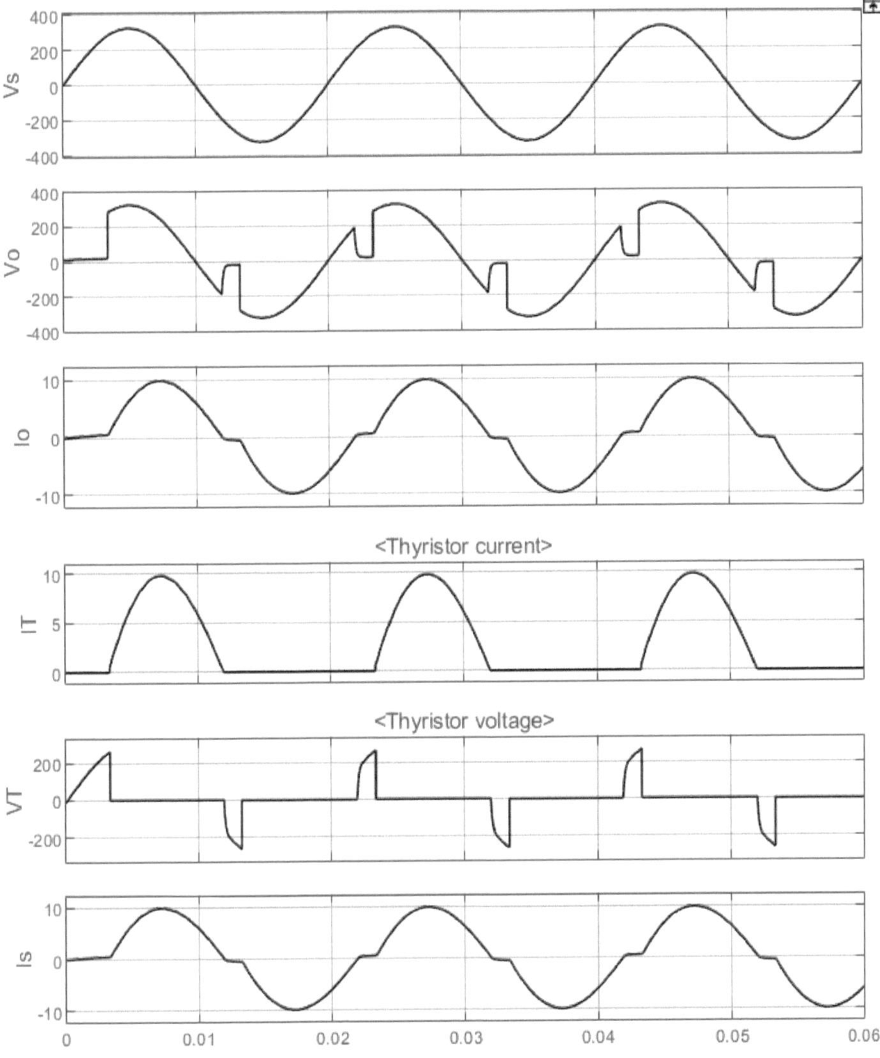

▲ **Fig. 6.5:** Voltage and current waveforms pertaining to phase A in a three-phase voltage controller

For the implementation of a three-phase voltage controller, six pulses are to be generated. Each pulse has to be separated by 60°. Voltage sensing of only one phase is needed for this purpose. Phase A voltage is sensed in the programme written below. Pin 3 is used for inputting the square wave synchronisation signal. Digital pins from 4 to 9 are used for generation triggering pulses for T1 to T6, respectively. The delay angle is to be inputted directly into the programme as the value of the variable alpha. The value is to be given as duration in microseconds. For a supply voltage at 50 Hz, 360° will be represented as 20,000 microseconds. An Arduino programme to achieve this is given below:

Programme 6.2

```
int val1;
int alpha=1333; //delay angle in microseconds for 30 degrees

void setup() {
  pinMode (3, INPUT); //for reading synchronizing square pulses
  pinMode(4,OUTPUT); // for triggering T1
  pinMode(5,OUTPUT); // for triggering T2
  pinMode(6,OUTPUT); // for triggering T3
  pinMode(7,OUTPUT); // for triggering T3
  pinMode(8,OUTPUT); // for triggering T3
  pinMode(9,OUTPUT); // for triggering T3
   }
void loop()
{

val1=digitalRead(3);

if ((val1)==1) //edge detection
{
  delayMicroseconds(alpha);// delay corresponding to alpha
  digitalWrite(4,HIGH);
  delay(1); //pulse width equivalent to 1 millisec
  digitalWrite(4,LOW);
  delayMicroseconds(2300); //remaining period in 60 degrees
```

```
  digitalWrite(5,HIGH);
  delay(1);
  digitalWrite(5,LOW);
  delayMicroseconds(2300);

  digitalWrite(6,HIGH);
  delay(1);
  digitalWrite(6,LOW);
  delayMicroseconds(2300);

  digitalWrite(7,HIGH);
  delay(1);
  digitalWrite(7,LOW);
  delayMicroseconds(2300);

  digitalWrite(8,HIGH);
  delay(1);
  digitalWrite(8,LOW);
  delayMicroseconds(2300);

  digitalWrite(9,HIGH);
  delay(1);
  digitalWrite(9,LOW);
}
}
```

6.3 Half-Controlled Voltage Controllers

If the range of control needed is limited to up to half of the supply voltage, one can go for half-controlled voltage controllers. In half-controlled voltage controllers, one thyristor in the thyristor pair in each phase is replaced with a diode. Hence, the control of either a positive or negative half-cycle is done while the full portion in the remaining half-cycle will appear at the output. The circuit diagram of a single-phase half-controlled voltage controller is shown in **Fig. 6.6**.

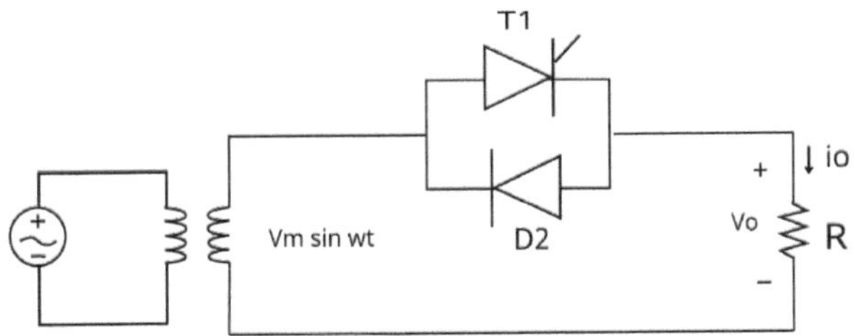

▲ **Fig. 6.6:** Single-phase half-controlled voltage controller

The output voltage waveform and the output current waveform along with the supply voltage, triggering signal and input current for a resistive load at a delay angle of 60° are shown in **Fig. 6.7**.

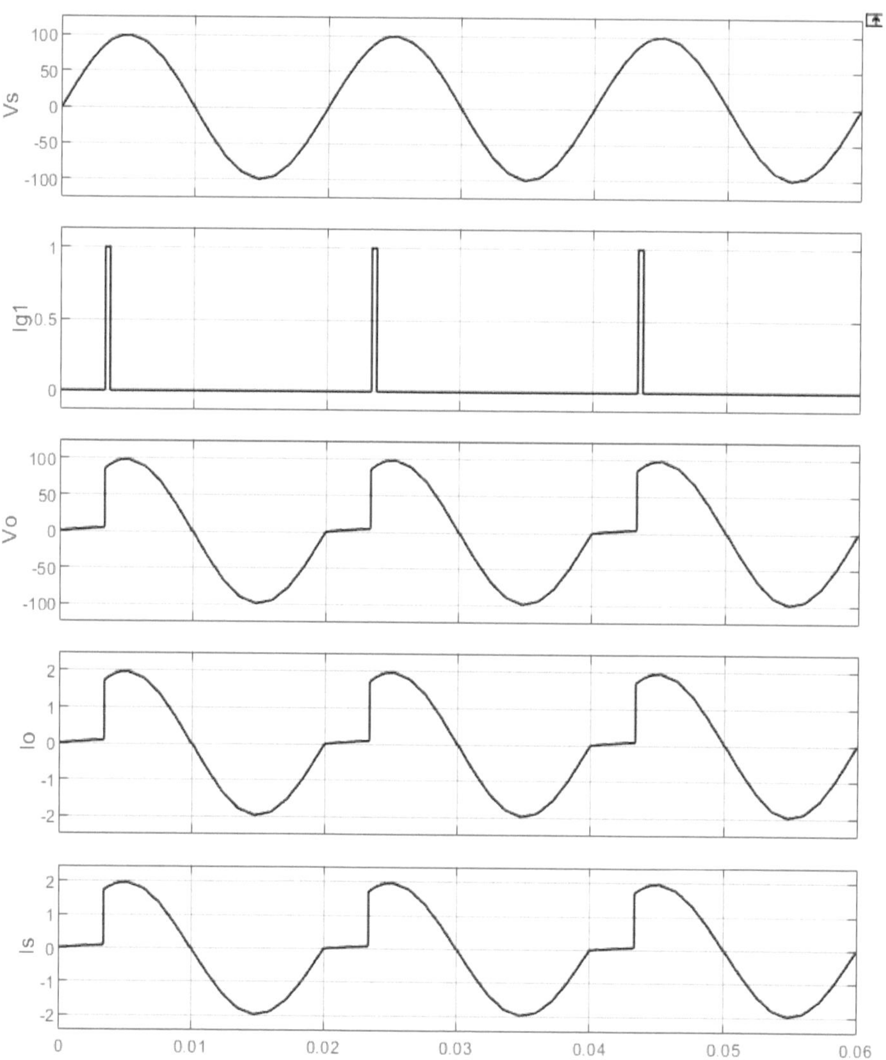

▲ **Fig. 6.7:** Waveforms of half-controlled voltage controller with R load for α=60°

The same waveforms with an RL load are shown in **Fig. 6.8**. It is to be noted that control action can be obtained only when the delay angle is more than the power factor angle of the load.

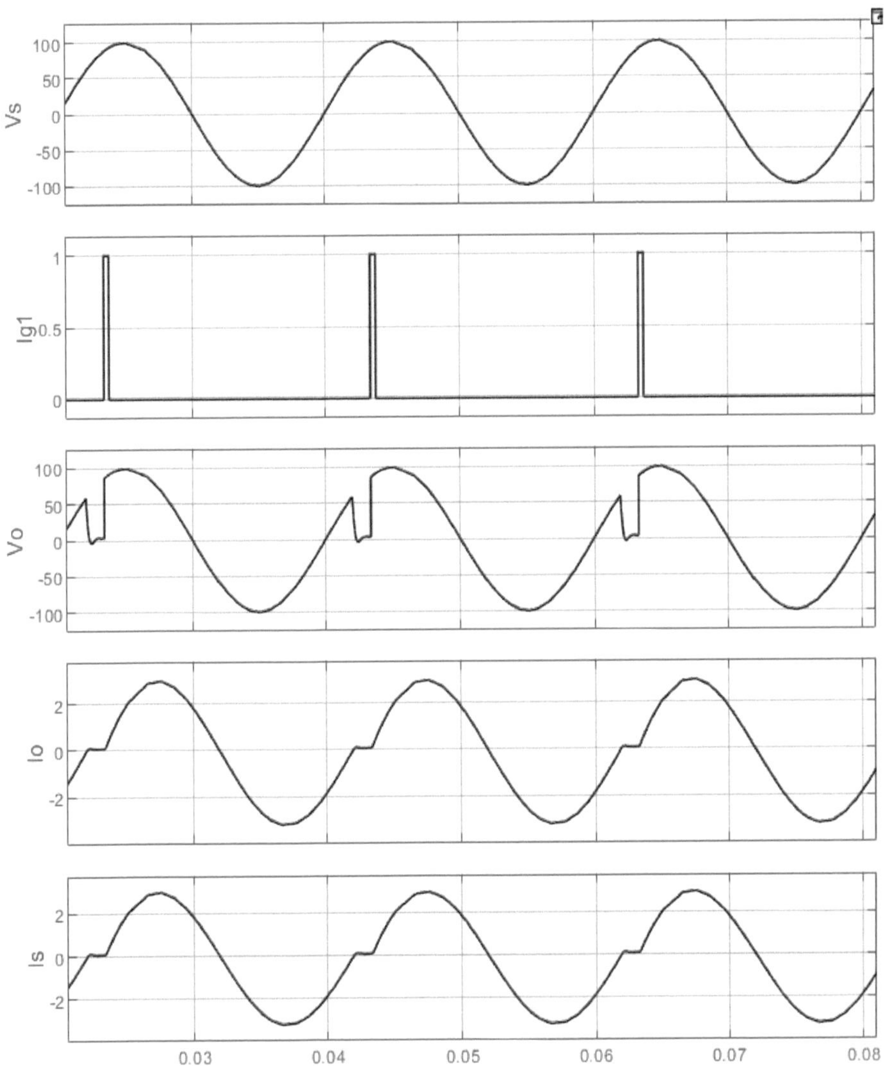

▲ **Fig. 6.7:** Waveforms of half-controlled voltage controller with RL load for α=60°

A half-controlled voltage controller can be implemented using the same programme written for a half-wave rectifier. It is needed to generate only one pulse delayed from the zero crossing point of the supply voltage. Synchronising signals generated in earlier circuits can be used here as well. Pin 3 is used for reading this signal and pin 4 is used for the generation of pulse for T1, in the programme given below:

Programme 6.3

```
int val1;
int alpha=3333; //delay for 60 degrees
void setup() {
  pinMode (3, OUTPUT);      //set 3 pin as output :
  pinMode(4,INPUT);
 }
void loop()
{

val1=digitalRead(4);

if ((val1)==1)
{
  delayMicroseconds(alpha);  // delay equivalent to alpha
  digitalWrite(3,HIGH);
  delay(1); //pulse width equivalent to 1 millisec
  digitalWrite(3,LOW);
  delay(7); //remaining period in positive cycle

 }

}
```

The circuit diagram of a three-phase half-controlled voltage controller is shown in **Fig. 6.8**. The thyristors which were used for conduction in the negative half-cycle in the case of a fully controlled voltage controller are replaced with diodes. The control of voltage is achieved in positive half-cycles alone. In negative half-cycles, the full voltage will appear across the load.

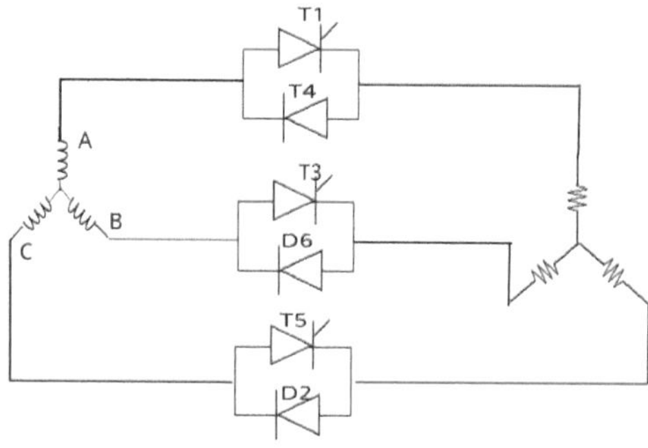

▲ **Fig. 6.8:** Three-phase half-controlled voltage controller

The voltage and current waves for an RL load at a delay angle of 60° for phase A are shown in **Fig. 6.9**.

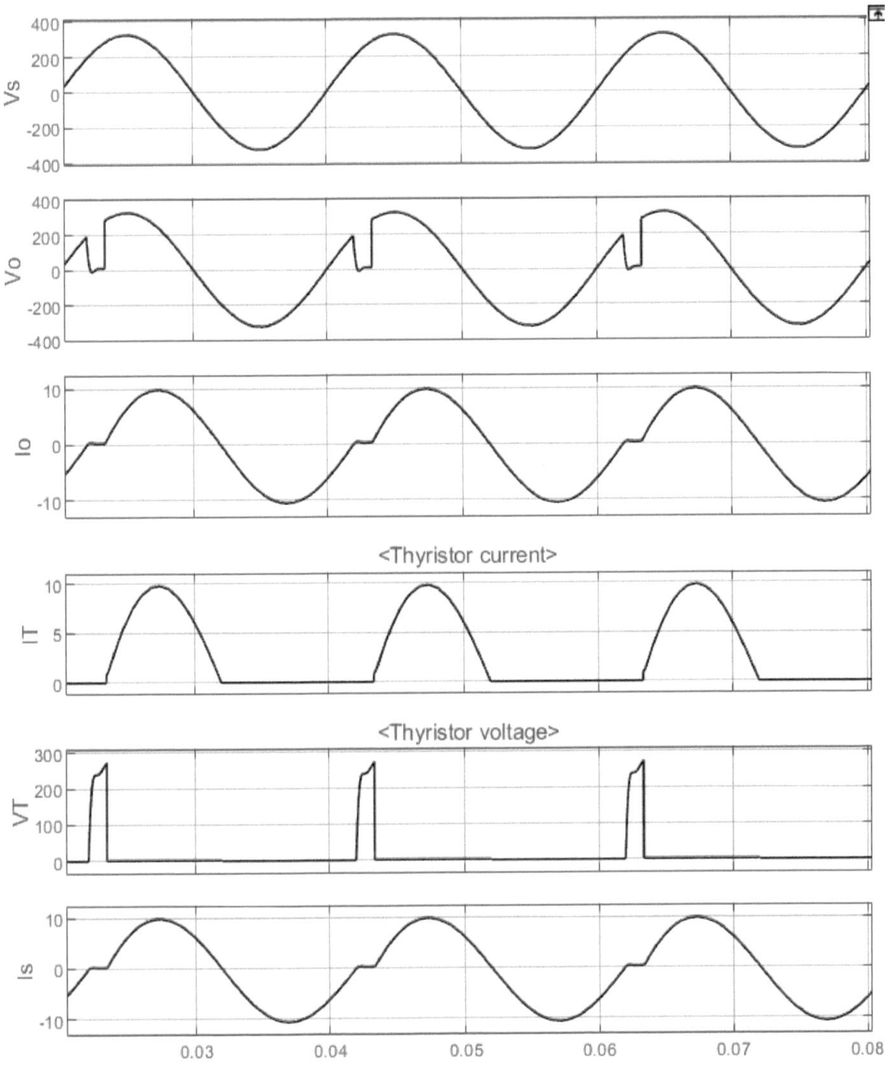

▲ **Fig. 6.9:** Voltage and current waveforms in a three-phase controller for an RL load working at a delay angle of 60°.

For generating the triggering pulses for a three-phase half-controlled voltage controller, three pulses separated by 120° are required from the zero crossing point of the phase A supply voltage. The programme used for the generation of a three-phase half-wave-controlled rectifier can be made use of. An Arduino programme which generates triggering pulses for T1, T3 and T5 at digital pins 4, 5 and 6, respectively, is shown below:

Program 6.4

```
int val1;
int alpha=3333; // value for 60 degree delay in microseconds

void setup() {
  pinMode (3, INPUT); //for reading synchronizing square pulses
  pinMode(4,OUTPUT); // for triggering T1
  pinMode(5,OUTPUT); // for triggering T2
  pinMode(6,OUTPUT); // for triggering T3
   }
void loop()
{
val1=digitalRead(3);

if ((val1)==1) //edge detection
{
  delayMicroseconds(alpha);  // delay corresponding to alpha
  digitalWrite(4,HIGH);
  delay(1); //pulse width equivalent to 1 millisec
  digitalWrite(4,LOW);
  delayMicroseconds(5666); //remaining period in 120 degrees


  digitalWrite(5,HIGH);
  delay(1);
  digitalWrite(5,LOW);
  delayMicroseconds(5666);
  digitalWrite(6,HIGH);
  delay(1);
  digitalWrite(6,LOW);
}
}
```

www.ingramcontent.com/pod-product-compliance
Ingram Content Group UK Ltd.
Pitfield, Milton Keynes, MK11 3LW, UK
UKHW041640190726
13854UKWH00006B/2609